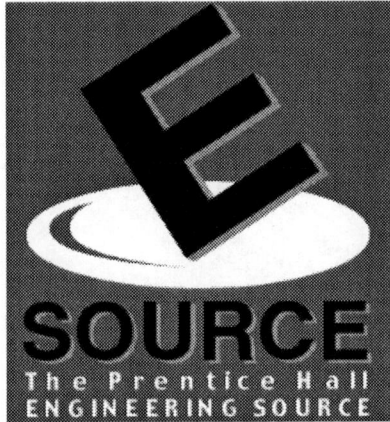

SOURCE
The Prentice Hall
ENGINEERING SOURCE

Introduction to Maple® 8

David I. Schwartz

Cornell University, Ithaca, NY

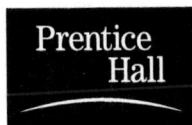

Prentice
Hall

Pearson Education, Inc.
Upper Saddle River, NJ 07458

Library of Congress Cataloging-in-Publication Data on file

Vice President and Editorial Director, ECS: *Marcia J. Horton*
Executive Editor: *Eric Svendsen*
Associate Editor: *Dee Bernhard*
Vice President and Director of Production and Manufacturing, ESM: *David W. Riccardi*
Executive Managing Editor: *Vince O'Brien*
Managing Editor: *David A. George*
Production Editor: *Scott Disanno*
Director of Creative Services: *Paul Belfanti*
Creative Director: *Carole Anson*
Art Director: *Jayne Conte*
Art Editor: *Greg Dulles*
Manufacturing Manager: *Trudy Pisciotti*
Manufacturing Buyer: *Lisa McDowell*
Marketing Manager: *Holly Stark*

Prentice Hall

© 2003 Pearson Education, Inc.
Pearson Education, Inc.
Upper Saddle River, New Jersey 07458

The author and publisher of this book have used their best efforts in preparing this book. These efforts include the development, research, and testing of the theories and programs to determine their effectiveness. The author and publisher make no warranty of any kind, expressed or implied, with regard to these programs or the documentation contained in this book. The author and publisher shall not be liable in any event for incidental or consequential damages in connection with, or arising out of, the furnishing, performance, or use of these programs.

Macintosh is a registered trademark of Apple Computer, Inc. Maple and Maple VIII are registered trademarks of Waterloo Maple, Inc. MATLAB is a registered trademark of The MathWorks, Inc. PostScript is a registered trademark of Adobe Systems, Inc. Unix is a registered trademark of the Open Group. Windows is a registered trademark of Microsoft, Inc.

Printed in the United States of America.

10 9 8 7 6 5 4 3 2 1

ISBN 0-13-032844-8

Pearson Education Ltd., *London*
Pearson Education Australia Pty. Ltd., *Sydney*
Pearson Education Singapore, Pte. Ltd.
Pearson Education North Asia Ltd., *Hong Kong*
Pearson Education Canada, Inc., *Toronto*
Pearson Educación de Mexico, S.A. de C.V.
Pearson Education—Japan, *Tokyo*
Pearson Education Malaysia, Pte. Ltd.
Pearson Education, *Upper Saddle River, New Jersey*

About ESource

ESource—The Prentice Hall Engineering Source—www.prenhall.com/esource

ESource—The Prentice Hall Engineering Source gives professors the power to harness the full potential of their text and their first-year engineering course. More than just a collection of books, ESource is a unique publishing system revolving around the ESource website—www.prenhall.com/esource. ESource enables you to put your stamp on your book just as you do your course. It lets you:

Control You choose exactly what chapters are in your book and in what order they appear. Of course, you can choose the entire book if you'd like and stay with the authors' original order.

Optimize Get the most from your book and your course. ESource lets you produce the optimal text for your students' needs.

Customize You can add your own material anywhere in your text's presentation, and your final product will arrive at your bookstore as a professionally formatted text. Of course, all titles in this series are available as stand-alone texts, or as bundles of two or more books sold at a discount. Contact your PH sales rep for discount information.

ESource ACCESS

Professors who choose to bundle two or more texts from the ESource series for their class, or use an ESource custom book, will be providing their students with an on-line library of intro engineering content—ESource Access. We've designed ESource ACCESS to provides students a flexible, searchable, on-line resource. Free access codes comes in bundles and custom books are valid for one year after initial log-on. Contact your PH sales rep for more information.

ESource Content

All the content in ESource was written by educators specifically for freshman/first-year students. Authors tried to strike a balanced level of presentation, an approach that was neither formulaic nor trivial, and one that did not focus too heavily on advanced topics that most introductory students do not encounter until later classes. Because many professors do not have extensive time to cover these topics in the classroom, authors prepared each text with the idea that many students would use it for self-instruction and independent study. Students should be able to use this content to learn the software tool or subject on their own.

While authors had the freedom to write texts in a style appropriate to their particular subject, all followed certain guidelines created to promote a consistency that makes students comfortable. Namely, every chapter opens with a clear set of **Objectives**, includes **Practice Boxes** throughout the chapter, and ends with a number of **Problems**, and a list of **Key Terms**. **Applications Boxes** are spread throughout the book with the intent of giving students a real-world perspective of engineering. **Success Boxes** provide the student with advice about college study skills, and help students avoid the common pitfalls of first-year students. In addition, this series contains an entire book titled ***Engineering Success*** by Peter Schiavone of the University of Alberta intended to expose students quickly to what it takes to be an engineering student.

Creating Your Book

Using ESource is simple. You preview the content either on-line or through examination copies of the books you can request on-line, from your PH sales rep, or by calling 1-800-526-0485. Create an on-line outline of the content you want, in the order you want, using ESource's simple interface. Insert your own material into the text flow. If you are not ready to order, ESource will save your work. You can come back at any time and change, re-arrange, or add more material to your creation. Once you're finished you'll automatically receive an ISBN. Give it to your bookstore and your book will arrive on their shelves four to six weeks after they order. Your custom desk copies with their instructor supplements will arrive at your address at the same time.

To learn more about this new system for creating the perfect textbook, go to www.prenhall.com/esource. You can either go through the on-line walk-through of how to create a book, or experiment yourself.

Supplements

Adopters of ESource receive an instructor's CD that contains professor and student code from the books in the series, as well as other instruction aides provided by authors. The website also holds approximately **350 Powerpoint transparencies** created by Jack Leifer of University of Kentucky-Paducah. Professors can either follow these transparencies as already pre-prepared lectures or use them as the basis for their own custom presentations.

Titles in the ESource Series

Design Concepts for Engineers, 2/e
0-13-093430-5
Mark Horenstein

Engineering Success, 2/e
0-13-041827-7
Peter Schiavone

Engineering Design and Problem Solving, 2E
ISBN 0-13-093399-6
Steven K. Howell

Exploring Engineering
ISBN 0-13-093442-9
Joe King

Engineering Ethics
0-13-784224-4
Charles B. Fleddermann

Introduction to Engineering Analysis
0-13-016733-9
Kirk D. Hagen

Introduction to Engineering Experimentation
0-13-032835-9
Ronald W. Larsen, John T. Sears, and Royce Wilkinson

Introduction to Mechanical Engineering
0-13-019640-1
Robert Rizza

Introduction to Electrical and Computer Engineering
0-13-033363-8
Charles B. Fleddermann and Martin Bradshaw

Introduction to MATLAB 6 — Update
0-13-140918-2
Delores Etter and David C. Kuncicky, with Douglas W. Hull

MATLAB Programming
0-13-013149-0
Delores Etter with David C. Kuncicky

Introduction to MATLAB
0-13-013149-0
Delores Etter with David C. Kuncicky

Introduction to Mathcad 2000
0-13-020007-7
Ronald W. Larsen

Introduction to Mathcad 11
0-13-008177-9
Ronald W. Larsen

Introduction to Maple 8
0-13-032844-8
David I. Schwartz

Mathematics Review
0-13-011501-0
Peter Schiavone

Power Programming with VBA/Excel
0-13-047377-4
Steven C. Chapra

Introduction to Excel 2002
0-13-008175-2
David C. Kuncicky

Engineering with Excel
ISBN 0-13-017696-6
Ronald W. Larsen

Introduction to Word 2002
0-13-008170-1
David C. Kuncicky

Introduction to PowerPoint 2002
0-13-008179-5
Jack Leifer

Graphics Concepts
0-13-030687-8
Richard M. Lueptow

Graphics Concepts with SolidWorks 2/e
0-13-140915-8
Richard M. Lueptow and Michael Minbiole

Graphics Concepts with Pro/ENGINEER
0-13-014154-2
Richard M. Lueptow, Jim Steger, and
Michael T. Snyder

Introduction to AutoCAD 2000
0-13-016732-0
Mark Dix and Paul Riley

Introduction to AutoCAD, R. 14
0-13-011001-9
Mark Dix and Paul Riley

Introduction to UNIX
0-13-095135-8
David I. Schwartz

Introduction to the Internet, 3/e
0-13-031355-6
Scott D. James

Introduction to Visual Basic 6.0
0-13-026813-5
David I. Schneider

Introduction to C
0-13-011854-0
Delores Etter

Introduction to C++
0-13-011855-9
Delores Etter

Introduction to FORTRAN 90
0-13-013146-6
Larry Nyhoff and Sanford Leestma

Introduction to Java
0-13-919416-9
Stephen J. Chapman

About the Authors

No project could ever come to pass without a group of authors who have the vision and the courage to turn a stack of blank paper into a book. The authors in this series, who worked diligently to produce their books, provide the building blocks of the series.

Martin D. Bradshaw was born in Pittsburg, KS in 1936, grew up in Kansas and the surrounding states of Arkansas and Missouri, graduating from Newton High School, Newton, KS in 1954. He received the B.S.E.E. and M.S.E.E. degrees from the University of Wichita in 1958 and 1961, respectively. A Ford Foundation fellowship at Carnegie Institute of Technology followed from 1961 to 1963 and he received the Ph.D. degree in electrical engineering in 1964. He spent his entire academic career with the Department of Electrical and Computer Engineering at the University of New Mexico (1961-1963 and 1991-1996). He served as the Assistant Dean for Special Programs with the UNM College of Engineering from 1974 to 1976 and as the Associate Chairman for the EECE Department from 1993 to 1996. During the period 1987-1991 he was a consultant with his own company, EE Problem Solvers. During 1978 he spent a sabbatical year with the State Electricity Commission of Victoria, Melbourne, Australia. From 1979 to 1981 he served an IPA assignment as a Project Officer at the U.S. Air Force Weapons Laboratory, Kirkland AFB, Albuquerque, NM. He has won numerous local, regional, and national teaching awards, including the George Westinghouse Award from the ASEE in 1973. He was awarded the IEEE Centennial Medal in 2000.

Acknowledgments: Dr. Bradshaw would like to acknowledge his late mother, who gave him a great love of reading and learning, and his father, who taught him to persist until the job is finished. The encouragement of his wife, Jo, and his six children is a never-ending inspiration.

Stephen J. Chapman received a B.S. degree in Electrical Engineering from Louisiana State University (1975), the M.S.E. degree in Electrical Engineering from the University of Central Florida (1979), and pursued further graduate studies at Rice University. Mr. Chapman is currently Manager of Technical Systems for British Aerospace Australia, in Melbourne, Australia. In this position, he provides technical direction and design authority for the work of younger engineers within the company. He also continues to teach at local universities on a part-time basis.

Mr. Chapman is a Senior Member of the Institute of Electrical and Electronics Engineers (and several of its component societies). He is also a member of the Association for Computing Machinery and the Institution of Engineers (Australia).

Steven C. Chapra presently holds the Louis Berger Chair for Computing and Engineering in the Civil and Environmental Engineering Department at Tufts University. Dr. Chapra received engineering degrees from Manhattan College and the University of Michigan. Before joining the faculty at Tufts, he taught at Texas A&M University, the University of Colorado, and Imperial College, London. His research interests focus on surface water-quality modeling and advanced computer applications in environmental engineering. He has published over 50 refereed journal articles, 20 software packages and 6 books. He has received a number of awards including the 1987 ASEE Merriam/Wiley Distinguished Author Award, the 1993 Rudolph Hering Medal, and teaching awards from Texas A&M, the University of Colorado, and the Association of Environmental Engineering and Science Professors.

Acknowledgments: To the Berger Family for their many contributions to engineering education. I would also like to thank David Clough for his friendship and insights, John Walkenbach for his wonderful books, and my colleague Lee Minardi and my students Kenny William, Robert Viesca and Jennifer Edelmann for their suggestions.

Mark Dix began working with AutoCAD in 1985 as a programmer for CAD Support Associates, Inc. He helped design a system for creating estimates and bills of material directly from AutoCAD drawing databases for use in the automated conveyor industry. This system became the basis for systems still widely in use today. In 1986 he began collaborating with Paul Riley to create AutoCAD training materials, combining Riley's background in industrial design and training with Dix's background in writing, curriculum development, and programming. Mr. Dix received the M.S. degree in education from the University of Massachusetts. He is currently the Director of Dearborn Academy High School in Arlington, Massachusetts.

Delores M. Etter is a Professor of Electrical and Computer Engineering at the University of Colorado. Dr. Etter was a faculty member at the University of New Mexico and also a Visiting Professor at Stanford University. Dr. Etter was responsible for the Freshman Engineering Program at the University of New Mexico and is active in the Integrated Teaching Laboratory at the University of Colorado. She was elected a Fellow of the Institute of Electrical and Electronics Engineers for her contributions to education and for her technical leadership in digital signal processing.

Charles B. Fleddermann is a professor in the Department of Electrical and Computer Engineering at the University of New Mexico in Albuquerque, New Mexico. All of his degrees are in electrical engineering: his Bachelor's degree from the University of Notre Dame, and the Master's and Ph.D. from the University of Illinois at Urbana-Champaign. Prof. Fleddermann developed an engineering ethics course for his department in response to the ABET requirement to incorporate ethics topics into the undergraduate engineering curriculum. *Engineering Ethics* was written as a vehicle for presenting ethical theory, analysis, and problem solving to engineering undergraduates in a concise and readily accessible way.

Acknowledgments: I would like to thank Profs. Charles Harris and Michael Rabins of Texas A & M University whose NSF sponsored workshops on engineering ethics got me started thinking in this field. Special thanks to my wife Liz, who proofread the manuscript for this book, provided many useful suggestions, and who helped me learn how to teach "soft" topics to engineers.

Kirk D. Hagen is a professor at Weber State University in Ogden, Utah. He has taught introductory-level engineering courses and upper-division thermal science courses at WSU since 1993. He received his B.S. degree in physics from Weber State College and his M.S. degree in mechanical engineering from Utah State University, after which he worked as a thermal designer/analyst in the aerospace and electronics industries. After several years of engineering practice, he resumed his formal education, earning his Ph.D. in mechanical engineering at the University of Utah. Hagen is the author of an undergraduate heat transfer text.

Mark N. Horenstein is a Professor in the Department of Electrical and Computer Engineering at Boston University. He has degrees in Electrical Engineering from M.I.T. and U.C. Berkeley and has been involved in teaching engineering design for the greater part of his academic career. He devised and developed the senior design project class taken by all electrical and computer engineering students at Boston University. In this class, the students work for a virtual engineering company developing products and systems for real-world engineering and social-service clients.

Acknowledgments: I would like to thank Prof. James Bethune, the architect of the Peak Performance event at Boston University, for his permission to highlight the competition in my text. Several of the ideas relating to brainstorming and teamwork were derived from a workshop on engineering design offered by Prof. Charles Lovas of Southern Methodist University. The principles of estimation were derived in part from a freshman engineering problem posed by Prof. Thomas Kincaid of Boston University.

Steven Howell is the Chairman and a Professor of Mechanical Engineering at Lawrence Technological University. Prior to joining LTU in 2001, Dr. Howell led a knowledge-based engineering project for Visteon Automotive Systems and taught computer-aided design classes for Ford Motor Company engineers. Dr. Howell also has a total of 15 years experience as an engineering faculty member at Northern Arizona University, the University of the Pacific, and the University of Zimbabwe. While at Northern Arizona University, he helped develop and implement an award-winning interdisciplinary series of design courses simulating a corporate engineering-design environment.

Douglas W. Hull is a graduate student in the Department of Mechanical Engineering at Carnegie Mellon University in Pittsburgh, Pennsylvania. He is the author of *Mastering Mechanics I Using Matlab 5*, and contributed to *Mechanics of Materials* by Bedford and Liechti. His research in the Sensor Based Planning lab involves motion planning for hyper-redundant manipulators, also known as serpentine robots.

Scott D. James is a staff lecturer at Kettering University (formerly GMI Engineering & Management Institute) in Flint, Michigan. He is currently pursuing a Ph.D. in Systems Engineering with an emphasis on software engineering and computer-integrated manufacturing. He chose teaching as a profession after several years in the computer industry. "I thought that it was really important to know what it was like outside of academia. I wanted to provide students with classes that

were up to date and provide the information that is really used and needed."

Acknowledgments: Scott would like to acknowledge his family for the time to work on the text and his students and peers at Kettering who offered helpful critiques of the materials that eventually became the book.

Joe King received the B.S. and M.S. degrees from the University of California at Davis. He is a Professor of Computer Engineering at the University of the Pacific, Stockton, CA, where he teaches courses in digital design, computer design, artificial intelligence, and computer networking. Since joining the UOP faculty, Professor King has spent yearlong sabbaticals teaching in Zimbabwe, Singapore, and Finland. A licensed engineer in the state of California, King's industrial experience includes major design projects with Lawrence Livermore National Laboratory, as well as independent consulting projects. Prof. King has had a number of books published with titles including *Matlab*, *MathCAD*, *Exploring Engineering*, and *Engineering and Society*.

David C. Kuncicky is a native Floridian. He earned his Baccalaureate in psychology, Master's in computer science, and Ph.D. in computer science from Florida State University. He has served as a faculty member in the Department of Electrical Engineering at the FAMU–FSU College of Engineering and the Department of Computer Science at Florida State University. He has taught computer science and computer engineering courses for over 15 years. He has published research in the areas of intelligent hybrid systems and neural networks. He is currently the Director of Engineering at Bioreason, Inc. in Sante Fe, New Mexico.

Acknowledgments: Thanks to Steffie and Helen for putting up with my late nights and long weekends at the computer. Finally, thanks to Susan Bassett for having faith in my abilities, and for providing continued tutelage and support.

Ron Larsen is a Professor of Chemical Engineering at Montana State University, and received his Ph.D. from the Pennsylvania State University. He was initially attracted to engineering by the challenges the profession offers, but also appreciates that engineering is a serving profession. Some of the greatest challenges he has faced while teaching have involved non-traditional teaching methods, including evening courses for practicing engineers and teaching through an interpreter at the Mongolian National University. These experiences have provided tremendous opportunities to learn new ways to communicate technical material. Dr. Larsen views modern software as one of the new tools that will radically alter the way engineers work, and his book *Introduction to MathCAD* was written to help young engineers prepare to meet the challenges of an ever-changing workplace.

Acknowledgments: To my students at Montana State University who have endured the rough drafts and typos, and who still allow me to experiment with their classes—my sincere thanks.

Sanford Leestma is a Professor of Mathematics and Computer Science at Calvin College, and received his Ph.D. from New Mexico State University. He has been the long-time co-author of successful textbooks on Fortran, Pascal, and data structures in Pascal. His current research interest are in the areas of algorithms and numerical computation.

Jack Leifer is an Assistant Professor in the Department of Mechanical Engineering at the University of Kentucky Extended Campus Program in Paducah, and was previously with the Department of Mathematical Sciences and Engineering at the University of South Carolina–Aiken. He received his Ph.D. in Mechanical Engineering from the University of Texas at Austin in December 1995. His current research interests include the modeling of sensors for manufacturing, and the use of Artificial Neural Networks to predict corrosion.

Acknowledgments: I'd like to thank my colleagues at USC–Aiken, especially Professors Mike May and Laurene Fausett, for their encouragement and feedback; and my parents, Felice and Morton Leifer, for being there and providing support (as always) as I completed this book.

Richard M. Lueptow is the Charles Deering McCormick Professor of Teaching Excellence and Associate Professor of Mechanical Engineering at Northwestern University. He is a native of Wisconsin and received his doctorate from the Massachusetts Institute of Technology in 1986. He teaches design, fluid mechanics, an spectral analysis techniques. Rich has an active research program on rotating filtration, Taylor Couette flow, granular flow, fire suppression, and acoustics. He has five patents and over 40 refereed journal and proceedings papers along with many other articles, abstracts, and presentations.

Acknowledgments: Thanks to my talented and hardworking co-authors as well as the many colleagues and students who took the tutorial for a "test drive." Special thanks to Mike Minbiole for his major contributions to Graphics Concepts with SolidWorks. Thanks also to Northwestern University for the time to work on a book. Most of all, thanks to my loving wife, Maiya, and my children, Hannah and Kyle, for supporting me in this endeavor. (Photo courtesy of Evanston Photographic Studios, Inc.)

Larry Nyhoff is a Professor of Mathematics and Computer Science at Calvin College. After doing bachelor's work at Calvin, and Master's work at Michigan, he received a Ph.D. from Michigan State and also did graduate work in computer science at Western Michigan. Dr. Nyhoff has taught at Calvin for the past 34 years—mathematics at first and computer science for the past several years.

Acknowledgments: We thank our families—Shar, Jeff, Dawn, Rebecca, Megan, Sara, Greg, Julie, Joshua, Derek, Tom, Joan; Marge, Michelle, Sandy, Lory, Michael—for being patient and understanding. We thank God for allowing us to write this text.

Paul Riley is an author, instructor, and designer specializing in graphics and design for multimedia. He is a founding partner of CAD Support Associates, a contract service and professional training organization for computer-aided design. His 15 years of business experience and 20 years of teaching experience are supported by degrees in education and computer science. Paul has taught AutoCAD at the University of Massachusetts at Lowell and is presently teaching AutoCAD at Mt. Ida College in Newton, Massachusetts. He has developed a program,

Computer-aided Design for Professionals that is highly regarded by corporate clients and has been an ongoing success since 1982.

Robert Rizza is an Assistant Professor of Mechanical Engineering at North Dakota State University, where he teaches courses in mechanics and computer-aided design. A native of Chicago, he received the Ph.D. degree from the Illinois Institute of Technology. He is also the author of *Getting Started with Pro/ENGINEER*. Dr. Rizza has worked on a diverse range of engineering projects including projects from the railroad, bioengineering, and aerospace industries. His current research interests include the fracture of composite materials, repair of cracked aircraft components, and loosening of prostheses.

Peter Schiavone is a professor and student advisor in the Department of Mechanical Engineering at the University of Alberta, Canada. He received his Ph.D. from the University of Strathclyde, U.K. in 1988. He has authored several books in the area of student academic success as well as numerous papers in international scientific research journals. Dr. Schiavone has worked in private industry in several different areas of engineering including aerospace and systems engineering. He founded the first Mathematics Resource Center at the University of Alberta, a unit designed specifically to teach new students the necessary *survival skills* in mathematics and the physical sciences required for success in first-year engineering. This led to the Students' Union Gold Key Award for outstanding contributions to the university. Dr. Schiavone lectures regularly to freshman engineering students and to new engineering professors on engineering success, in particular about maximizing students' academic performance.

Acknowledgements: Thanks to Richard Felder for being such an inspiration; to my wife Linda for sharing my dreams and believing in me; and to Francesca and Antonio for putting up with Dad when working on the text.

David I. Schneider holds an A.B. degree from Oberlin College and a Ph.D. degree in Mathematics from MIT. He has taught for 34 years, primarily at the University of Maryland. Dr. Schneider has authored 28 books, with one-half of them computer programming books. He has developed three customized software packages that are supplied as supplements to over 55 mathematics textbooks. His involvement with computers dates back to 1962, when he programmed a special purpose computer at MIT's Lincoln Laboratory to correct errors in a communications system.

David I. Schwartz is an Assistant Professor in the Computer Science Department at Cornell University and earned his B.S., M.S., and Ph.D. degrees in Civil Engineering from State University of New York at Buffalo. Throughout his graduate studies, Schwartz combined principles of computer science to applications of civil engineering. He became interested in helping students learn how to apply software tools for solving a variety of engineering problems. He teaches his students to learn incrementally and practice frequently to gain the maturity to tackle other subjects. In his spare time, Schwartz plays drums in a variety of bands.

Acknowledgments: I dedicate my books to my family, friends, and students who all helped in so many ways.

Many thanks go to the schools of Civil Engineering and Engineering & Applied Science at State University of New York at Buffalo where I originally developed and tested my UNIX and Maple books. I greatly appreciate the opportunity to explore my goals and all the help from everyone at the Computer Science Department at Cornell.

John T. Sears received the Ph.D. degree from Princeton University. Currently, he is a Professor and the head of the Department of Chemical Engineering at Montana State University. After leaving Princeton he worked in research at Brookhaven National Laboratory and Esso Research and Engineering, until he took a position at West Virginia University. He came to MSU in 1982, where he has served as the Director of the College of Engineering Minority Program and Interim Director for BioFilm Engineering. Prof. Sears has written a book on air pollution and economic development, and over 45 articles in engineering and engineering education.

Michael T. Snyder is President of Internet startup Appointments123.com. He is a native of Chicago, and he received his Bachelor of Science degree in Mechanical Engineering from the University of Notre Dame. Mike also graduated with honors from Northwestern University's Kellogg Graduate School of Management in 1999 with his Masters of Management degree. Before Appointments123.com, Mike was a mechanical engineer in new product development for Motorola Cellular and Acco Office Products. He has received four patents for his mechanical design work. "Pro/ENGINEER was an invaluable design tool for me, and I am glad to help students learn the basics of Pro/ENGINEER."

Acknowledgments: Thanks to Rich Lueptow and Jim Steger for inviting me to be a part of this great project. Of course, thanks to my wife Gretchen for her support in my various projects.

Jim Steger is currently Chief Technical Officer and cofounder of an Internet applications company. He graduated with a Bachelor of Science degree in Mechanical Engineering from Northwestern University. His prior work included mechanical engineering assignments at Motorola and Acco Brands. At Motorola, Jim worked on part design for two-way radios and was one of the lead mechanical engineers on a cellular phone product line. At Acco Brands, Jim was the sole engineer on numerous office product designs. His Worx stapler has won design awards in the United States and in Europe. Jim has been a Pro/ENGINEER user for over six years.

Acknowledgments: Many thanks to my co-authors, especially Rich Lueptow for his leadership on this project. I would also like to thank my family for their continuous support.

Royce Wilkinson received his undergraduate degree in chemistry from Rose-Hulman Institute of Technology in 1991 and the Ph.D. degree in chemistry from Montana State University in 1998 with research in natural product isolation from fungi. He currently resides in Bozeman, MT and is involved in HIV drug research. His research interests center on biological molecules and their interactions in the search for pharmaceutical advances.

Reviewers

We would like to thank everyone who has reviewed texts in this series.

ESource Reviewers

Christopher Rowe, *Vanderbilt University*
Steve Yurgartis, *Clarkson University*
Heidi A. Diefes-Dux, *Purdue University*
Howard Silver, *Fairleigh Dickenson University*
Jean C. Malzahn Kampe, *Virginia Polytechnic Institute and State University*
Malcolm Heimer, *Florida International University*
Stanley Reeves, *Auburn University*
John Demel, *Ohio State University*
Shahnam Navee, *Georgia Southern University*
Heshem Shaalem, *Georgia Southern University*
Terry L. Kohutek, *Texas A & M University*
Liz Rozell, *Bakersfield College*
Mary C. Lynch, *University of Florida*
Ted Pawlicki, *University of Rochester*
James N. Jensen, *SUNY at Buffalo*
Tom Horton, *University of Virginia*
Eileen Young, *Bristol Community College*
James D. Nelson, *Louisiana Tech University*
Jerry Dunn, *Texas Tech University*
Howard M. Fulmer, *Villanova UniversityBerkeley*
Naeem Abdurrahman, *University of Texas, Austin*
Stephen Allan, *Utah State University*
Anil Bajaj, *Purdue University*
Grant Baker, *University of Alaska–Anchorage*
William Beckwith, *Clemson University*
Haym Benaroya, *Rutgers University*
John Biddle, *California State Polytechnic University*
Tom Bledsaw, *ITT Technical Institute*
Fred Boadu, *Duke University*
Tom Bryson, *University of Missouri, Rolla*
Ramzi Bualuan, *University of Notre Dame*
Dan Budny, *Purdue University*
Betty Burr, *University of Houston*
Dale Calkins, *University of Washington*
Harish Cherukuri, *University of North Carolina –Charlotte*
Arthur Clausing, *University of Illinois*

Barry Crittendon, *Virginia Polytechnic and State University*
James Devine, *University of South Florida*
Ron Eaglin, *University of Central Florida*
Dale Elifrits, *University of Missouri, Rolla*
Patrick Fitzhorn, *Colorado State University*
Susan Freeman, *Northeastern University*
Frank Gerlitz, *Washtenaw College*
Frank Gerlitz, *Washtenaw Community College*
John Glover, *University of Houston*
John Graham, *University of North Carolina–Charlotte*
Ashish Gupta, *SUNY at Buffalo*
Otto Gygax, *Oregon State University*
Malcom Heimer, *Florida International University*
Donald Herling, *Oregon State University*
Thomas Hill, *SUNY at Buffalo*
A.S. Hodel, *Auburn University*
James N. Jensen, *SUNY at Buffalo*
Vern Johnson, *University of Arizona*
Autar Kaw, *University of South Florida*
Kathleen Kitto, *Western Washington University*
Kenneth Klika, *University of Akron*
Terry L. Kohutek, *Texas A&M University*
Melvin J. Maron, *University of Louisville*
Robert Montgomery, *Purdue University*
Mark Nagurka, *Marquette University*
Romarathnam Narasimhan, *University of Miami*
Soronadi Nnaji, *Florida A&M University*
Sheila O'Connor, *Wichita State University*
Michael Peshkin, *Northwestern University*
Dr. John Ray, *University of Memphis*
Larry Richards, *University of Virginia*
Marc H. Richman, *Brown University*
Randy Shih, *Oregon Institute of Technology*
Avi Singhal, *Arizona State University*
Tim Sykes, *Houston Community College*
Neil R. Thompson, *University of Waterloo*
Dr. Raman Menon Unnikrishnan, *Rochester Institute of Technology*
Michael S. Wells, *Tennessee Tech University*
Joseph Wujek, *University of California, Berkeley*
Edward Young, *University of South Carolina*
Garry Young, *Oklahoma State University*
Mandochehr Zoghi, *University of Dayton*

Contents

1

Introduction

1.1 A BEGINNING

Why should you learn Maple? Perhaps a teacher or colleague has told you that Maple is a great program for solving hard problems, or perhaps you simply are required to learn it. Teachers often expect students to believe in Maple's importance on faith. Throughout this text, I hope to *demonstrate* Maple's power and elegance to you. For now, though, read through this chapter to get a "taste" for Maple.

OBJECTIVES

After reading this chapter, you should be able to

- Relate engineering and science to problem solving
- Compare mathematics and computers to tools
- Understand how Maple operates
- Explain the problem-solving process
- Describe the notations employed in this text
- Identify elements of the Maple workspace
- Access on-line help
- Create, open, save, and print worksheets

1.1.1 Problems

Maple involves problem solving. What distinguishes *engineering* and *science* problems from other problems? Certainly not all tasks require an engineer or scientist! Loosely defined, scientists *develop* theoretical concepts of the physical world, and engineers then *apply* these theories to create physical systems, devices such as lightbulbs and toenail clippers. Yes, physics helps to keep our nails trimmed!

1.1.2 Tools

How do people construct devices? Usually, they do so with **tools**. Tools assist with thinking, analysis, testing, and making things that solve problems. Tools include physical *devices* and abstract *methods*. Physical tools, like machinery, extend human physical capabilities. Abstract tools, like algorithms, enhance problem solving.

1.1.3 Modeling

Models are tools that analyze and predict physical behaviors. Science and engineering models often combine features of physical and abstract tools. Using algorithms and equations, for example, manufacturers often build prototypes before mass marketing goods for testing after designing. Well-devised models reasonably and efficiently predict accurate behaviors. The abstract model provides a design, whereas the prototype provides a reality check to see if the device really works.

1.1.4 Experimentation

Experimentation yields good models. For instance, consider a building and the idealized model of it, as shown in Figure 1.1. Over many years, laboratory testing and theoretical development have provided engineers with idealized mathematical models that assist in structural design. Continued testing between idealized and actual behaviors improves both understanding and development of better models.[1]

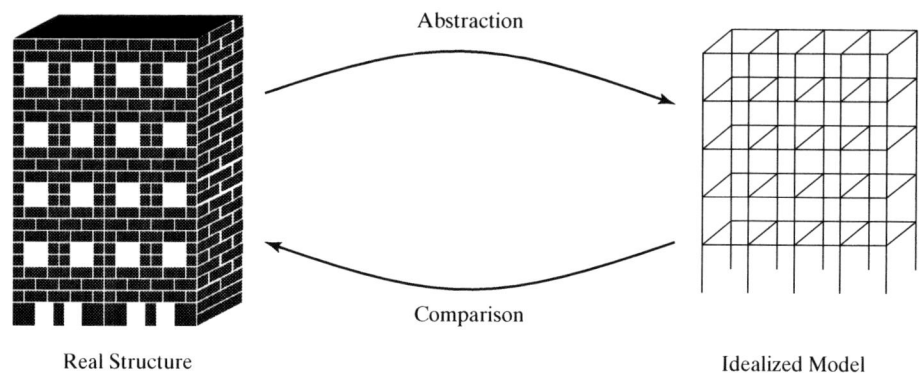

Figure 1.1 Basic Modeling

---

[1]Can buildings have prototypes, like other mass-marketed goods? Imagine having to destroy buildings in succession until constructing one that didn't topple! Abstract models alleviate the need for such wanton destructiveness.

1.1.5   Mathematics

Mathematics provides abstract models. Scientists and engineers represent physical processes with equations. Mathematical models usually involve a combination of symbolic and numerical features:

- A ***symbolic model*** is *qualitative*, because it represents a general equation. For instance, the equation $y = mx + b$ represents a symbolic mathematical model, where the letters $y, m, x,$ and b represent different numerical values. Qualitative analysis involves studying how values change as variables increase or decrease.
- A ***numerical model*** is *quantitative*, because it substitutes specific numerical values for the terms in a symbolic model. For instance, using values $m = 1, x = 2,$ and $b = 3$ in the equation $y = mx + b$, produces a quantitative result of $y = 5$.

1.1.6   Computers

Tools such as ***computers*** assist in complex mathematical analysis and other burdensome tasks. Hardware and software make up the "body" and "mind," respectively, of a computer. Hardware consists of the physical devices; software provides the "intelligence" for computations and communication. So what makes software a tool? As with any other tool,[2] software helps people solve problems.

1.2   MAPLE

Maple provides a powerful software tool used primarily for symbolic analysis, but it has recently strengthened its numerical abilities as well. This section provides a brief overview of Maple's capabilities and structure.

1.2.1   Computer Algebra System

Maple implements a ***computer algebra system*** (CAS) for symbolic mathematics. A CAS provides commands for manipulating and deriving symbolic equations from other equations. A software "engine" inside a CAS uses *analytical methods* to find exact answers, often in symbolic form. Maple employs numerical methods as well and now implements many numerical routines with NAG (Numerical Algorithms Group).

1.2.2   History of Maple

Computer-based, symbolic analysis was introduced in the late 1960s. In case you're curious where the name *Waterloo Maple* came from, the Symbolic Computation Group at the University of Waterloo in Canada started Maple as a project in 1980. Eventually, Waterloo Maple, Inc., commercially released Maple, which has remained popular in both academic and commercial environments ever since.

1.2.3   Why Maple?

Maple is very useful for a number of reasons:

- It performs symbolic and numerical analysis for a wide variety of tasks.
- It offers a rich library of thousands of functions.
- It helps you construct customized programs by using Maple commands.
- It provides access to other popular software packages and Internet technology.

---

[2]Application software implements mathematical models for analysis and design. You should understand a program's assumptions and limitations, as you would with any model! Always check your output with simpler "textbook" models.

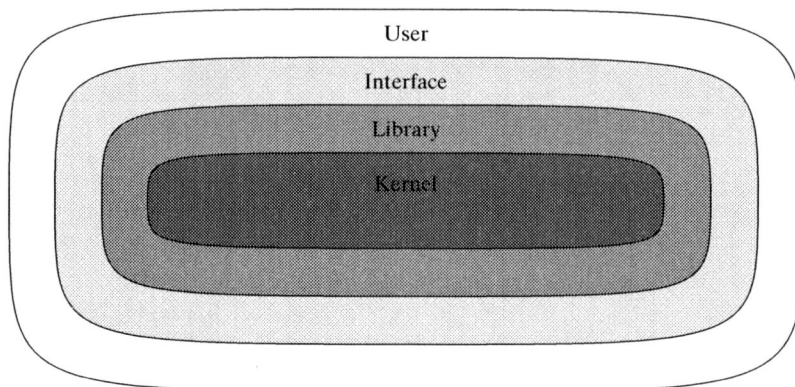

Figure 1.2 Maple Structure

1.2.4 How Does Maple Work?

The **kernel**, **library**, and an **interface** make up the structure of Maple, as shown in Figure 1.2:

- *Kernel*: About 10% of Maple functions make up the Maple kernel, the core command group. These commands interpret and process input, evaluate numerical quantities, and manage memory. Maple provides both *shared* and *parallel* kernels, as discussed in Chapter 3.
- *Library*: About 90% of Maple's commands are called *functions*, which are written in Maple code and kernel operations. Functions automatically load when starting Maple, but some functions are stored in library packages that require special loading procedures, as discussed in Appendix B.
- *Interface*: The user communicates with Maple through the interface. Maple interprets commands and evaluates the desired procedures. Although Maple provides a text-based interface, this text focuses on the **graphical user interface** (GUI) version of Maple, which uses menus and windows.
- *User*: That's you!

1.3 USING MAPLE

This text covers *Maple 8*. A few important distinctions with previous versions are noted throughout the text. You should carefully review the material in this section to help with learning the material and doing problems.

1.3.1 Problem-Solving Methodology

Each chapter in this book applies Maple for solving a variety of problems. As you work through these problems, try to follow a general problem-solving methodology, such as the one illustrated in Figure 1.3. In general, you should break problems down into smaller, more manageable sub-problems. When a step fails, go back to the previous step and iterate for better solutions. You might wish to think of this process as a more refined "trial-and-error" system.

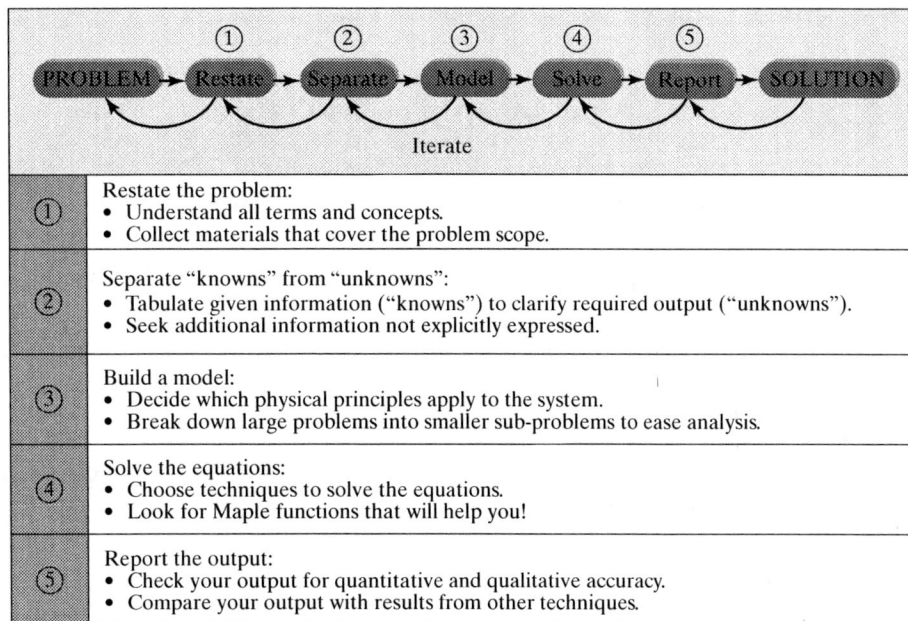

Figure 1.3 A Problem-Solving Methodology

1.3.2 Notation

Before attempting anything, thoroughly review the notation in Table 1.1. Computer manuals and textbooks typically are awash with such notation, so practicing now will help for future study. I suggest that you mark this page because you'll need to refer it while getting used to this text.

TABLE 1.1 Notation

Notation	Description
key terms	These ***terms*** involve important concepts and are shown in boldface italics. There is a list of key terms at the end of each chapter.
`input`	`input` and `commands` are shown in boldface, fixed-width font. For example, you would enter the mathematical expression sin5 as `sin(5)`.
`arguments`	Many Maple commands take *arguments*, which are values that can vary. Arguments are shown in boldface italicized fixed-width font. Never type the literal word "***argument***"! Always type the value of "***argument***" instead.
output and *equations*	GUI Maple reports most *output* in italics. Why? Many textbooks show mathematical expressions and equations in italics. For example, entering `sin(x)` in Maple would produce the output sin(x).
computer	When you make a mistake, Maple sometimes produces output in fixed-width font to report your blunder.
text	You may also add plain text in Maple. Although Maple allows different fonts, this book shows plain text in this font.
commentary	My *comments* are shown in this small, italic font. Never enter these comments into Maple commands!
↵	Press the Enter or Return key on your keyboard.
Menu1 → Menu2	Select Menu items by pointing and clicking the left mouse button. For instance, the sequence Edit→Execute→Worksheet instructs you to select Edit. You may now select from the submenu. Point your mouse on Execute. Finally, click on Worksheet. You can also press the Alt key along with the underlined letter to select a particular item on the first in a sequence of menu selections. Do not press Alt again to activate a submenu.
Topic…Subtopic	Point and click your mouse inside the Help Browser for help on specific topics.

1.4   MAPLE ENVIRONMENT

This section provides a brief overview of Maple's GUI environment. You should try the suggested *Practice!* problems after reading the following material.

1.4.1   Starting Maple

Maple provides two basic environments:

- Command Line, as shown in Figure 1.4. Consider using this text-based version when you have become more experienced with Maple or need to run only a few commands.
- GUI, as shown with annotations in Figure 1.5. In this text, we focus on the GUI version.

To start the GUI version of Maple,

- Windows or MacIntosh: Click the Maple icon or select the appropriate menu item.
- Unix or Linux: Enter **xmaple** or **maple -x** at the shell prompt.

Refer to Waterloo Maple's website http://www.maplesoft.com/ for a complete list of supported operating systems.

```
dis@mobius% maple

     |\^/|       Maple 8 (SUN SPARC SOLARIS)
 ._|\|   |/|_. Copyright (c) 2002 by Waterloo Maple Inc.
  \  MAPLE  /  All rights reserved. Maple is a registered trademark of
  <____ ____>  Waterloo Maple Inc.
       |        Type ? for help.

> Enter commands here. Exit with quit,done, or stop.
```

Figure 1.4   Command-Line Maple Interface

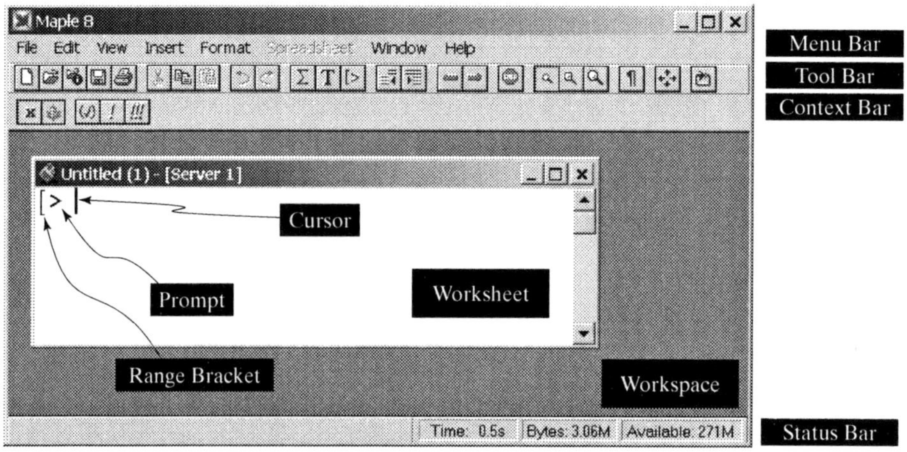

Figure 1.5   Maple Window

TABLE 1.2 Elements of Maple GUI

Feature	Description	Reference
Menu Bar	Select commands with your mouse or the Alt key.	`?menus`
Tool Bar	Click these icons for shortcuts. These icons change based on the tasks you perform, like accessing help or plots.	`?toolbar`
Context Bar	Change input and output formats, halt execution, and edit commands.	`?contextbar`
Title Bar	Click the Title Bar icons for resizing, iconifying, and closing a worksheet.	`?glossary`
Worksheet	Create documents that communicate with Maple.	`?worksheet`
Workspace	Open multiple worksheets and other Maple windows.	`?windowmenu`

1.4.2   GUI Interface

You're about to take your first steps through Maple. Start up the GUI version of Maple and compare the window that pops up with Figure 1.5. Look at the components of the window, which are summarized in Table 1.2.

1.4.3   Item Selection

Maple uses two different methods to access menus with the mouse:

- A *left click* means that you press and release the left mouse button (Windows and Unix) or the sole mouse button (MacIntosh) to select menus, worksheets, and items inside worksheets.
- A *right click* means that you press and release the right mouse button (Windows and Unix) or press the Option key (MacIntosh) to perform actions from the context-sensitive menu. (See Getting Started...Menus...Context Sensitive Menus...).
- A *double* or *triple click* means that you rapidly left-click two or three times, respectively.

In general, to select a menu, move the mouse until the cursor points over the item and left-click, unless a right click is specified.

PRACTICE!

1. Which appendix in this book has solutions to Practice problems?
2. Locate and activate Maple on your system.
3. Which version of Maple do you use?

1.5   HELP!

This section will help you become familiar with Maple's on-line help. A "seasoned" Maple user tends to rely on Maple's on-line help to answer questions, but many Help features assist new and inexperienced users.

1.5.1 Help Menu

To investigate Maple's Help system, left-click on the <u>H</u>elp menu using your mouse. When first using Maple,

- Select <u>H</u>elp→**Preferences...** to open a window for saving options that affect how Maple operates.
- Click the **Balloon Help** option.
- Click **Apply Globally** to store the option for all sessions.

Little balloons with helpful information will appear when you touch any icon and menu option. You might eventually find these balloons irritating, but until you know them all, I recommend leaving the balloon help on.

1.5.2 Tutorials

New Maple users should select <u>H</u>elp→<u>N</u>ew User's Tour and <u>H</u>elp→<u>I</u>ntroduction for an overview of Maple features.

1.5.3 Help Browser

Maple has an excellent tool for finding help on commands called the Help Browser, as shown in Figure 1.6. You can open the Help Browser with the following actions:

- Select <u>H</u>elp→<u>U</u>sing Help or other menu selections like <u>H</u>elp→<u>I</u>ntroduction.
- Type **?** or **??** at the Maple prompt (>), and press Enter or Return (↵).
- Find help on commands using **?command** or a text or topic search under <u>H</u>elp.

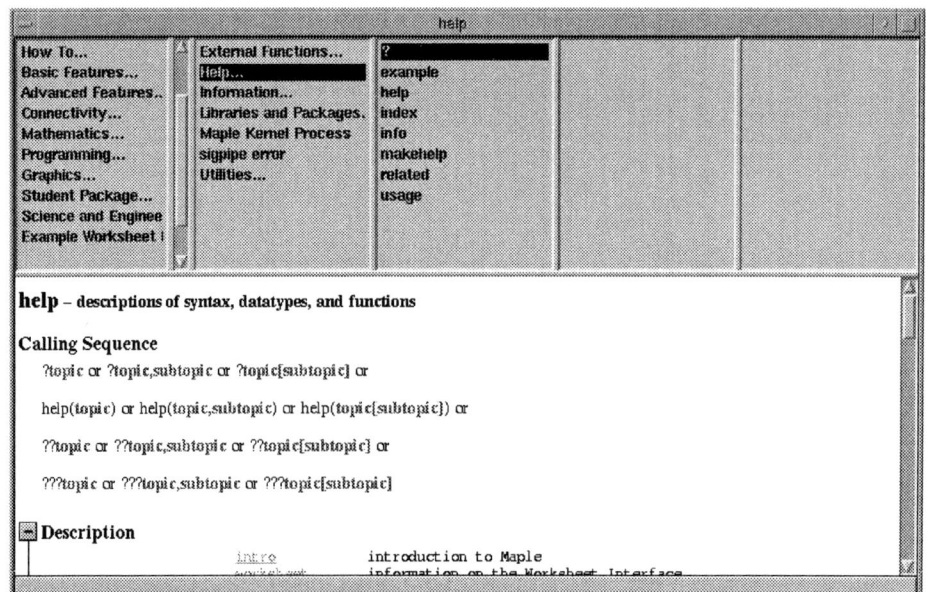

Figure 1.6 Help Browser

The Help Browser organizes many of Maple's resources in categories called *topics*. Typically, topics describe commands with help on syntax, usage, and examples. Left-click topics (listed as **Topic...**) in the left column to open *subtopics* that will appear to the right of the topics. To move forward and backward to other topics, click Help's Tool Bar icons.

1.5.4   Sections and Subsections

Help topics typically organize information with *sections*, indicated by the icons ⊞ and ⊟. Sections collect text, execution groups (inputs and outputs), and other sections. *Subsections* are sections contained within other sections.

- Left-click the *open-box* icon ⊞ to expand a section.
- Left-click the *collapse-box* icon ⊟ to collapse a section.

1.5.5   Hyperlinks

Many topics and other keywords are provided as *hyperlinks*. Hyperlinks are active text that "jump" to other topics or display information. Maple typically displays hyperlinks as colored, underlined text. Left-click a hyperlink once to activate it.

1.5.6   Command Line Help

Regardless of interface, you can enter **?topic** at the Maple prompt (>) for help on **topic**:

- Type a question mark (?) followed by **command**.
- Press the Enter or Return key (↵).

Maple pops up the Help Browser with the appropriate Help page loaded. If you are uncertain of the topic name, which is usually a command name, just type one or two letters. Maple will display all topic names that start with those letters.

1.5.7   Command Line Completion

If you are unsure of a command, Maple can attempt to complete it. Type a few letters and then press the F6 key, or select **Edit**→**Complete Command**. Maple will try to finish the name. See **?completecommand** for more information.

1.5.8   Topic Search

Select **Help**→**Topic Search...** to search for topics such as command names. You should type a few letters of the topic name in the box next to **Topic**. As long as the **Auto-search** feature is activated in the search window, Maple will automatically search for matching topics as you type the letters. For instance, try searching for **dis**, as shown in Figure 1.7. Maple will suggest topics like **disassemble** and **discont**. Double-click the topic names to investigate Maple's suggestions. If you select **discont**, Maple will pop open the Help Browser with the **discont** topic page loaded.

1.5.9   Text Search

If you are uncertain of a topic or command name, try a text search with **Help**→**Full Text Search...**. In the box next to **Word(s)**, type the words for which you are searching and press the **Search** button. Maple will suggest topics that may match your inquiry. Double-click a topic to open the Help page associated with the topic.

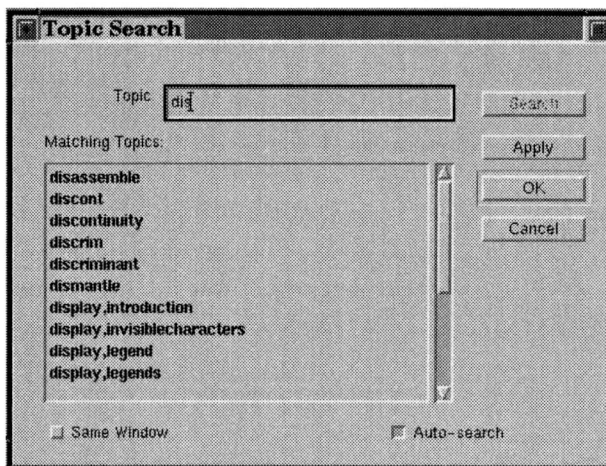

Figure 1.7   Topic Search

1.5.10   Examples

Maple provides many example worksheets that cover a wide variety of topics and applications. For on-line examples, enter **?examples[index]** at the prompt. For a much larger directory, refer to the *Maple Application Center* at Maple's website.

PRACTICE!

4. What Help topic in Maple contains the word *suck*? (Yes, there really is one.)
5. Activate Maple's Balloon Help feature. Point your mouse on an icon and on a menu. What happens?
6. Find Maple's on-line glossary inside the Help Browser.

1.6   THE MAPLE WORKSHEET

The Maple ***worksheet*** provides your main interface for communicating and working with Maple. In a way, a Maple worksheet resembles a document you would create with a word processor. At the prompt, you may enter commands, find help, and write text to create documents that contain commentary and the results of computations. This chapter explores many of these features as described in **?worksheet**, **?worksheet[interface]**, and **?worksheet[management]**.

1.6.1   Creating a Worksheet

By default, Maple starts with an empty worksheet. To work on a different worksheet, create a new worksheet by selecting the icon 🗋 or menu File→New. During a session, Maple labels each worksheet as "Untitled [number]" on the Title Bar in ascending order according to "newness." See How To...Create A Worksheet for more information.

1.6.2   Workspace

Select Window options to organize and rearrange worksheets and all other Maple windows inside your workspace. For instance, try opening a few worksheets inside a Maple workspace. You may then select <u>W</u>indow→<u>T</u>ile to display all windows next to one another in a tile pattern.

1.6.3   Moving Around Your Worksheet

Generally, you may move around your worksheet by doing one of the following actions:

- Point and click the mouse inside the worksheet.
- Press the arrow keys to change the position of the cursor.

Avoid moving the cursor with the Backspace, Del, and Enter (↵) keys. In the Help Browser, consult How To...Navigate... for further information on other ways to move around the worksheet. Eventually, you should investigate keyboard shortcuts for your operating system in **?hotmac**, **?hotunix**, or **?hotwin**.

1.6.4   Editing

To perform editing functions, such as copying, cutting, and pasting, use the mouse to highlight portions of your worksheet:

- Move the mouse to the portion of the text you wish to edit.
- Left-click the mouse and hold the mouse button.
- Drag the mouse over the portion you are editing.
- Select <u>E</u>dit or the Tool Bar icons for the editing operations.

You can also delete items by selecting them with your left mouse button and pressing the Del and Backspace keys. Consult How To... Copy, Cut, Move or Paste..., **?editmenu**, **?documenting**, and **?toolbar** for more information.

1.6.5   Saving

Always save your work! You never know when the system might crash. Select the icon 💾 or the menu <u>F</u>ile→**Save As**... to name and save worksheets:

- <u>F</u>olders or Directory: Change folders and directories by double-clicking the folder name.
- File <u>n</u>ame: Type a name that ends with **.mws**, like **hw1.mws**.
- Save file as <u>t</u>ype or Filetype: Leave the type on Maple Worksheet. Worksheets saved in *Maple Worksheet format* to preserve the entire contents in a format that you open directly into Maple.

When your chosen file name satisfies you, click OK. The new name will appear on the left side of the Title Bar on the worksheet. Once you have named your worksheet, frequently select <u>F</u>ile→Save to save your work. For more information, consult How To... Save... and **?worksheet**, **managing**, **saving**.

1.6.6   Exporting Worksheets

You can save your work to different formats by selecting <u>F</u>ile→<u>E</u>xport As.... If you prefer, you can also select <u>F</u>ile→<u>S</u>ave As... and then save the worksheet to another type of file. Note that exporting a worksheet to another format only exports the current version of your work. You have to export your worksheet again if you wish to incorporate changes inside the saved file. Some common formats include the following:

- HTML: Convert your worksheet to HTML for the World Wide Web. See **?xprt2html** for more details.

- Maple Text: A text-based representation of the Maple worksheet. The command-line version of Maple typically saves in this format. In general, however, you should stick to using MWS, instead.

Consult How To... Export... and **?worksheet**, **managing**, **export** for more information about these and other formats.

1.6.7 Opening Worksheets

To load a previously saved worksheet, you *open* the worksheet by selecting the icon 🖻 or menu File→Open. The window that pops up resembles the "Save As" window. Select a directory by clicking Folders or Directory. The Open File window automatically lists files with **mws** extensions. Click an existing file or type the name of a file under File name. Click OK to load the worksheet. For information, consult How To... Export... and **?worksheet**, **managing**, **export**.

1.6.8 Printing

If you attempt to print an MWS file without using Maple, your printer will produce garbled printouts. Instead, use Maple to print your worksheet:

- First, try File→Print Preview... to view a file before printing it.
- Next, select the icon 🖨 or menu File→Print....

The Print window gives you two choices for printing:

- File: Click Print to file for Windows/Macs and Output to File for Unix. A checkmark (✔) or indented diamond ◈ will appear in the box to the right. Click OK. Maple will then prompt you to enter a file name. Choose a name, like **something.ps**. Maple then creates a PostScript file containing the contents of your worksheet.
- Hardcopy: Ensure that the Print to file box has no checkmark (✔). Then, click OK. Unix users should select Print Command and then enter **lpr -Pprinter** or **lp -dprinter** in the box to the right.

For more information, consult How To... Print....

1.6.9 PostScript

PostScript (PS) is another kind of text-based formatting that creates files that print on virtually every kind of printer. The first line in a PS file typically starts with %!PS-Adobe. Beware of the difference between PostScript and Maple's MWS files! Maple cannot open PostScript files, so always retain both PostScript and MWS versions of your worksheets. Remember that you should never directly print a file stored in MWS format unless you use Maple's File→Print... command sequence!

1.6.10 Closing Windows

Refer to How To... Close... in the Help Browser. Note that closing a window, like a worksheet, does not terminate your Maple session.

1.6.11 Quitting Maple

To end your entire Maple session, enter File→Exit. Maple will then prompt you to save unsaved worksheets.

PRACTICE!

7. Create a new worksheet.
8. Save the worksheet to a file called **temp1.mws**. What file type should you pick for the file? What does the Title Bar for the worksheet now display?
9. Print the worksheet to a file called **temp1.ps**.
10. Close the worksheet called **temp1.mws**.
11. Open the worksheet that you stored in **temp1.mws**. Can you load the file called **temp1.ps**? Why or why not?
12. Exit from Maple.

PROFESSIONAL SUCCESS: CHECK YOUR RESULTS!

Never blindly trust computers. Consider the following story: For one assignment, my students had to estimate a person's weight, given a random sampling of height and weight data. Using statistical formulas, the students determined models to predict individual weights from the data. Unfortunately, some students did not check their work or think about what the results meant. Some reports estimated individual weights as averaging over a relatively uncommon 500 pounds. A few students even determined that some people weigh −80,000 pounds! Worse, many of these students left no explanation for these highly dubious results.

How can you avoid making such mistakes? Respect the adage, "garbage in, garbage out." Often, a misplaced number or character can yield rather bizarre answers. Also, perhaps, a model might break the assumptions required by the computer's implementation. Computer-software bugs present even more insidious problems. Input data and output might appear reasonable, but inside lurks an error. Protect yourself by trying these tips:

- Think about the numbers that your program outputs.
- Run multiple test cases.
- Investigate Maple's **?verify** library.
- Try different software packages and compare results among them.
- Test your results with manual methods found in textbooks whenever possible.
- Ask yourself, "Do the answers make physical sense?"

Remember that you bear the responsibility to produce safe, reliable solutions.

KEY TERMS

analytical method	computer	computer algebra system (CAS)
engineering	experimentation	graphical user interface (GUI)
Maple	Maple interface	Maple library
Maple kernel	Maple worksheet format (MWS)	mathematics
models	numerical model	PostScript (PS)
science	symbolic model	tools
worksheet		

SUMMARY

- A variety of tools help solve science and engineering problems.
- Tools include both physical devices and abstract methods.
- Computer programs implement models derived from scientific principles and experimentation to ease analysis and improve accuracy.
- Computer algebra systems provide both symbolic and numerical analysis.

- Maple is a computer algebra system.
- Three basic components form Maple: the kernel, libraries, and interface.
- You may search for help using Maple's Help Browser, accessed via the <u>H</u>elp menu or the prompt (?).
- You should save worksheets in Maple's own MWS format, but you may not print those files. To print a worksheet, you should use <u>F</u>ile→<u>P</u>rint....

2

Maple Overview

2.1   MAPLE INPUT AND OUTPUT

Maple provides a robust tool for mathematics, programming, and plotting. In general, you will fill Maple worksheets with a collection of commands that Maple will act upon. This section introduces basic methods of entering commands that will generate results in your worksheets.

OBJECTIVES

After reading this chapter, you should be able to

- Use the Maple graphical user interface
- Describe the components of a Maple worksheet
- Enter input to produce output
- Collect input and output in an execution group
- Perform basic computations and plotting in Maple
- Use palettes and context-sensitive menus

No matter how exciting and engaging Maple, or any material, may be, nitty-gritty details might drag you down. But learning need not bore you! Pique your interest and motivation with the following tips:

- Familiarize yourself with the text: Successful professionals never know everything—they just know where to look. Quickly flip though all chapters and appendices of the book. Look for handy features, like solutions to problems and reference tables. You should also compare your course syllabus with the text's organization.

- Learn the notation: Thoroughly understand all notation! Otherwise, you will always be lost.

- Set goals: Carefully read your assignments and text, but never try to learn everything at once. Attempt only a few portions at a time. Taking small steps will improve your retention and prevent burnout.

- Practice reading: Learning requires both understanding and practice. Study the text by skimming the assigned readings a few times so you gain a general understanding of the new material. Then, carefully read smaller portions and practice the demonstrated examples. To stay focused, periodically remind yourself of chapter objectives.

- Practice implementing: Follow the suggested examples in the book. Always try to understand the reasoning behind each exercise and predict the possible output before attempting the commands. Read the text above and below each command for explanations of syntax and behavior.

- Be Patient! Maple and other software sometimes confound students, initially. Hence, you should expect some frustration that will subside eventually. But, with enough practice and patience, you can learn Maple and become as good at it as anyone.

2.1.1 Maple Input

Move the cursor after a right angle bracket (>), which is the *prompt*. You will enter commands **Maple Input** which consists of commands and mathematical expressions:

[>   *This is the Maple prompt that you should never type! Enter Maple input to the right of it.*

Maple displays all executable input in red by default. Consult **?contextmaplein** in **?worksheet[reference]** for descriptions of Context Bar icons that can help you enter input.

2.1.2 Input Modes

Maple provides two input methods—***Maple Notation*** and ***Standard Math***—as demonstrated in Table 2.1.

- *Maple Notation*: Input is one-dimensional. Maple reproduces keyboard character input for mathematical expressions. Terminate this input by typing a semicolon (**;**) or colon (**:**). To process the input, press Enter or Return (↵).

TABLE 2.1   Maple Input Modes

Input	Prompt	Example
Maple Notation	[>|	**(a + b)/c**
Standard Math	[>?	$\dfrac{a + b}{c}$

- *Standard Math*: Input is two-dimensional. Maple converts keyboard character input into typeset, textbook notation. To terminate and process input, press ↵ twice.

Note that you should never terminate help commands, like **?command**, with a semi-colon or colon.

2.1.3 Choosing Input Modes

This text employs Maple Notation. You may switch input modes for a particular expression by first highlighting that expression with your mouse. You can then

- Either select the notation toggle-icon ☒ or
- Click the right mouse button to select or deselect Standard Math.

You can also set a default mode with File→Preferences...:

- Choose the I/O Display tab.
- Choose Input Display and Output Display options. To be consistent with this text, select Maple Notation for input and Standard Math Notation for output.
- Click Apply Globally.

Consult **?insertmenu** and **?worksheet[expressions]** for more information.

2.1.4 Maple Output

After you press Enter (↵) after typing Maple input, Maple executes the input and produces results called **Maple output**. Most Maple output is shown in blue by default, though Maple produces some output (typically, single characters which do mathematical operations) in black. To see an example of output produced by Maple, compute the sum $1 + 1$ using the Maple Notation **1+1**:

Step 1: Maple Input and Output

> **1+1;** ↵   *Red Input: Type* **1+1;** *and then press Enter. Evaluate the expression* **1+1**.

2   *Blue Output: Maple produces the output* 2.

To help manipulate output with Context Bar icons, consult **?contextmapleout**.

2.1.5 Error Messages

Inevitably, everyone confronts complaints from a computer in the form of error messages. For instance, what happens if you forget a semicolon or colon? Maple will complain by reporting an error message as output.

Step 2: Error Messages

> **1+1** ↵   *Enter* **1+1** *without a terminating semicolon or colon.*

>   *The cursor will blink on this line.*

Warning, premature end of input   *Maple reports a common error message.*

What should you do? Don't worry, and don't get frustrated. Trust me, there is a solution. You really should satisfy Maple's demands by fixing the error. You may do this without deleting the mistake:

- Place the cursor *after* the input **1+1** or at the next prompt (>).
- Type a semicolon and then press ↵.

Maple will show the correct output 2. Now, Maple is "happy" once again. You should also try Maple's automatic correction mechanism by clicking 🖉 , which sometimes helps.

2.1.6 Multiple Inputs

Semicolons and colons terminate input command statements:

- A semicolon (**;**) signals the end of input. Maple evaluates the input and *shows* the output.
- A colon (**:**) tells Maple not only to perform the computations, but to *suppress* the results.

To place separate commands on the same input line, you may use semicolons and colons. For example, try evaluating $1 + 1$ and $1 + 2$ on the same input line, but only show the results of $1 + 2$:

Step 3: Multiple Inputs

```
> 1+1: 1+2;↵              Compute 1+1 but do not show output. Next, compute and show 1+2.
            3              The colon suppresses the output of 1+1.
```

Yes, Maple really did compute $1 + 1$, even though you do not see output. When you learn about assignments, reasons for suppressing output become more apparent. (Yes, a little faith is required when learning software.) Consult **?separator** for more information.

PRACTICE!

1. Enter $\frac{1}{2}$ in both Maple Notation and Standard Math. (Hint: Try the forward slash key to create the fraction.)
2. Ensure Maple Notation is the default input option.
3. Evaluate $1 + 2 + 3$ with Maple.
4. Evaluate $1 + 1$ and 1×2 on the same input line. Repeat this problem, but suppress the output of $1 + 1$ the second time.

2.2   DOCUMENTATION

Why should you document your work? Today, your worksheet might seem entirely clear, but what about a year from now? And what about colleagues/graders/bosses that need to review your work? Not only should your work be accurate; it should be *clear* as well! This section helps you with basic elements of adding commentary to your worksheets. Also, Maple provides many word-processing functions. For more information, consult **?worksheet[documenting]**, **?comments**, or try the menu options in <u>F</u>ormat.

2.2.1 Text

You may document your work with **_text_**, which is composed of inert, nonexecutable input.[1] Maple displays text in black, by default, though you may change the appearance as discussed in **?styles**. To enter text, try one of the following three methods which produce the same effect:

- Move the cursor one space to the left of the prompt, as shown in Figure 2.1. The worksheet is now primed for text entry, so start typing.
- Select Insert→Text Input and type your text.
- Click the text icon **T** and type your text.

When finished, click ▷ to get a Maple **prompt**. If you use the first method, you may also press ↵. See **?instext** for more methods of text entry. Also, note that while in text mode, the Context Bar changes with new fields and icons.

Now you will enter text, using the methods described earlier. Each time you press ↵, a prompt will appear:

Step 4: Entering Text

I am text!↵	*Move the cursor left one space. Enter this line.*
I am text, <u>too</u>!↵	*Move the cursor left one space. Enter this line.*
>	*The prompt reappears each time you press ↵.*

To underline the word "too,"

- Highlight the word "too" with your mouse.
- Click the underline icon **U** in the Context Bar while still in text mode.

Refer to **?contexttext** for more information.

2.2.2 Comments

You may also document your work by mixing commentary and Maple input. To create a comment on an input line, use the number sign symbol (**#**). Maple ignores all input following the **#**:

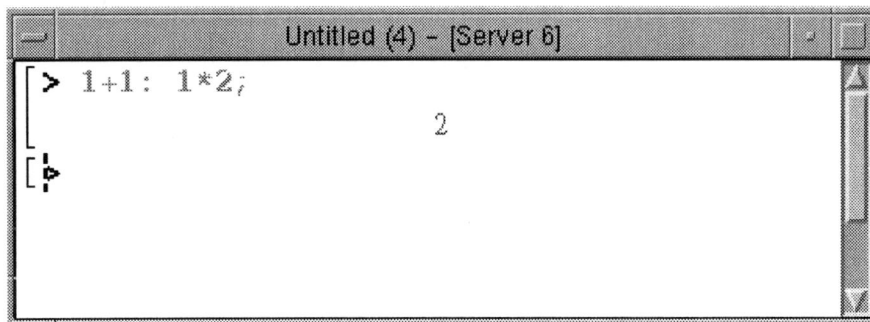

Figure 2.1   Entering Text by Moving the Cursor Left of the Prompt

---

[1]Beware that *text* differs from *Maple text*, a format in which you may save entire worksheets.

Step 5: Maple Comments

```
> 1*2; # multiply 1 and 2↵
                         2
```
Evaluate 1×2 *and include a comment.*
Maple produced output 2 *but ignored the comment.*

Consult **?comment** for more information.

2.2.3   Inert Math

You may also convert any executable, Maple input into ***inert input***, a nonexecutable form of Maple input shown in black. To do so, highlight the Maple input shown in red, and do one of the following:

- Click the leaf icon 🍁 , or
- Click the right mouse button and deselect **Maple Input**.

To return input to an active, executable state, select 🍁 . Or right-click your mouse, followed by selecting **Maple Input**.

PRACTICE!

5. Evaluate $1 + 1$ along with the comment *I am a comment* on the same input line.
6. How do you convert inert input into executable Standard Math?
7. Can you convert text to Maple Input? If so, how?

2.3   EXECUTION GROUPS

Have you noticed how Maple collects input and output together with range brackets on the left side of a worksheet? Maple is trying to help you document your work. This section introduces further techniques for helping you produce clear reports by collecting input, text, and output together.

2.3.1   Paragraphs

Maple considers each line of text, input, and output as separate *paragraphs*:

- Add lines by selecting <u>I</u>nsert→<u>P</u>aragraph followed by either <u>B</u>efore or <u>A</u>fter.
- Delete lines with <u>E</u>dit→<u>D</u>elete Paragraph or the Del and Backspace keys.

Lines that contain graphical items like spreadsheets and plots are also paragraphs. To display where Maple terminates paragraphs, select the ¶ icon or <u>V</u>iew→**Show Invisible Characters**. Maple will identify individual paragraphs using the symbol ¶ and show other normally invisible characters.

2.3.2   Creating Execution Groups

Execution groups collect paragraphs together. Execution groups can consist of one or more paragraphs. Typically, input lines are followed by output lines inside one execution group. Range brackets connect input and output paragraphs, as shown in Figure 2.2. Common tasks you may perform with execution groups include the following:

- Selecting <u>V</u>iew→**Show Group Ranges** to show or remove brackets.
- Pressing ↵ or the icon ▷ to create a new execution group.
- Selecting <u>I</u>nsert→**Execution Group** followed by <u>B</u>efore Cursor or <u>A</u>fter Cursor to insert a new group before or after the cursor.

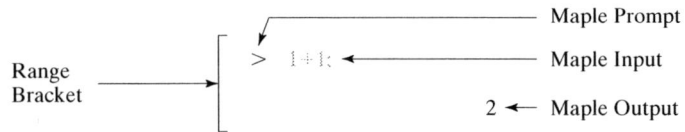

Figure 2.2   Execution Group

2.3.3   Joining Execution Groups

I strongly suggest that you practice "playing" with software as a way to learn. Thus, this section demonstrates steps that help you create execution groups to organize your work. You will create two individual execution groups and then connect them to create the worksheet shown in Figure 2.3.

Inside a new execution group, insert a paragraph of text and input. If you enter text with T select Insert→Maple Input to insert a new paragraph of input:

Step 6: Create First Group

I am text.↵	*First, add a line of text.*
> **1+1;**↵	*Next, enter input.*
2	*Maple generates and includes output in the group.*

At the next prompt that appears, enter **1+2**:

Step 7: Create Second Group

> **1+2;**↵	*Enter input.*
3	*Maple generates and includes output in the group.*

To join the groups in Steps 6 and 7,

- Place the cursor in the first execution group (Step 6), which has the input **1+1**.
- Select Edit→Split or Join→Join Execution Groups.

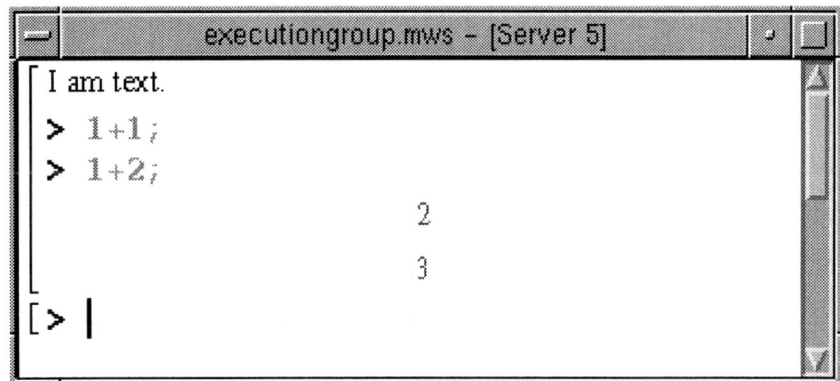

Figure 2.3   Example Execution Group

Now, you will see the following execution group:

Step 8: Joined First and Second Groups

I am text.	*This execution group is joined from the previous two groups.*
> **1+1;**	*This input is from Step 6.*
2	*This output is from Step 6.*
> **1+2;**	*This input is from Step 7.*
3	*This output is from Step 7.*

To make this group resemble Figure 2.3, click anywhere inside the group and press ↵ to reorganize the input and output lines. Maple places all input paragraphs together and evaluates each input in sequence, from top to bottom:

Step 9: Reorganize Input and Output Lines

I am text.
> **1+1;**
> **1+2;**

2
3

Refer to **?joingroup** and **?executiongroups** for more fun things you can do with execution groups.

PRACTICE!

8. Select <u>V</u>iew→Show <u>I</u>nvisible Characters or the ¶ icon. What do you see? Now turn off the invisible-character display.
9. Create an execution group with three paragraphs of text and no Maple input.
10. Evaluate $1 + 1$, $1 - 1$, and 1×2 on separate paragraphs in the same execution group.

2.4 MAPLE FEATURES

If you have not taken Maple's on-line tour yet, perhaps you have only witnessed Maple's arithmetic skills. But, Maple is much more than a grandiose calculator! Maple offers you powerful tools for solving complicated problems. Before solving those problems, however, you need to continue developing your skills as a Maple user. So, try the features in this section before continuing further. You should also consult **?worksheet** or Basic Features... for on-line overviews.

2.4.1 Case Sensitivity

Maple distinguishes between upper and lower case letters: Maple is ***case sensitive***! For instance, never enter **a** when you mean **A**!

2.4.2 Expressions

A mathematical ***expression***, like $2x + y$, combines numbers and symbols with *operators*. You have already used basic arithmetic operators **+**, **−**, *****, and **/**. Maple denotes symbols, like the variables x and y, as *names*. Chapter 3 reviews names and operators. Refer to **?manipulation** for tips on selecting and editing expressions.

2.4.3   Functions

Maple *functions* help build expressions and have the structure **func(arg1, arg2, ...)**:

- **func** is the function name
- **arg1**, **arg2**, ..., are arguments that the function acts upon.

Suppose you want to take the square root of a number using Maple. Surely, someone working for Maple has already programmed this for you. So, how do you find the function? Try the following problems to find out.

PRACTICE!

> 11.   Perform a search for the text *square root*.

Once you know the name of the function **func**, find help by entering **?func** at the Maple prompt, as discussed in Section 1.5.6. For example, enter **?sqrt** for help on **sqrt**:

Step 10: Find Help on Function

> **?sqrt**↵   *Find help on* **sqrt**. *Maple opens a window that discusses* **sqrt**.

To find the square root of an expression **expr**, enter **sqrt(expr)**:

Step 11: Using a Function

> **sqrt(4);**↵   *Find the square root of 4.*
>   2   *Maple succeeded! For more fun, enter* **sqrt(x^2)**.

Later chapters introduce many helpful functions for a variety of tasks. Appendix H summarizes all the functions used in this text. For further information on how Maple stores and accesses functions, refer to Appendix B.

2.4.4   Assignments

An ***assignment*** stores an expression "inside" a Maple variable name. Maple assignments have the syntax **name := expr**, where **name** stores **expr** inside Maple's memory. For instance, if I told you to store the value 72 in a variable called x, you would enter **x:=72** at a worksheet prompt. Then, every time Maple sees x, Maple really uses 72. So, a name resembles a *variable*, which acts as a placeholder for another expression. When entering assignments,

- Always put a colon (**:**) before the equals sign (**=**) with no space between them!
- Never use a lone equals sign (**=**) for assignments!

Try the following assignment, where you store the expression $1 + 1$ using variable *Var*:

Step 12: Introducing Assignments

> **Var := 1+1;** ⏎
$$Var:=2$$

*Compute **1+1** and store the value "inside" Var.*
Maple reports your assignment.

Maple first performs computations required by the expression $1 + 1$ before performing the assignment. (Later on you will discover why.) Maple now considers all future occurrences of *Var* as the value 2. Thus, you can check the value of *Var* by entering the name as input:

Step 13: Checking Assigned Values

> **Var;** ⏎
$$2$$

*What value does **Var** contain?*
*Maple reports the value of **Var**.*

A new worksheet opened during a session might actually "know" previously assigned variables, as discussed in Chapter 3. So, to erase assignments from Maple's memory, enter **restart** at the prompt in the worksheet in which you are working.

PRACTICE!

12. Store the value 2 in x. Store the value 1 in X.
13. Check the values of x and X. Why should x and X report different output?
14. Now, enter **restart**. Check the values of x and X. What happened?

2.4.5 Graphics

A picture says a thousand words ... when modeling equations, consider using graphs, or as Maple calls them, ***plots***. When analyzing a model, a plot demonstrates qualitative behaviors not readily apparent in the mathematics.

Maple provides an excellent tool for generating plots of expressions. For instance, try plotting the equation, $y = f(x) = 2x + 3$ on $0 \le x \le 5$. To plot y as a function of x, use a function of the form **plot(*expr*, *range*, *options*)** with the following arguments:

- ***expr:*** enter the expression $2x + 3$ as **2*x+3**. If you previously entered **y:=2*x+3**, you may enter **y**, instead. But, never enter **y=2*x+3** because **plot** cannot handle expressions that include the equal sign.
- ***range:*** enter the interval of the independent variable $0 \le x \le 5$ as the range **x=0..5**.
- ***option:*** display a title using the option **title="plot title"**.

Step 14: Basic Plotting

> **plot(2*x+3, x=0..5, title="Test Plot");** ⏎ *Plot $f(x) = 2x+3$ along $0 \le x \le 5$.*

Test Plot

(plot: a straight line rising from about 3 at x=0 to about 13 at x=5; y-axis labeled 4, 6, 8, 10, 12; x-axis labeled 0, 1, 2, 3, 4, 5; x-axis title "x")

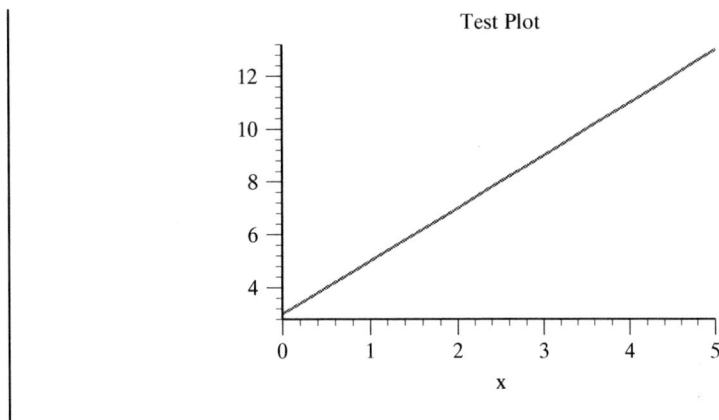

Maple's default setup has plots appear in an execution group. To place plots in a separate window, select the menu Option→Plot Display→Window. Review Chapter 8 and Graphics… to explore more of Maple's plotting features.

PRACTICE!

15.   Assign the expression $2\sqrt{x}$ to y.
16.   Plot y for $0 \le x \le 100$. Title your plot "Hello, I am a plot."

2.4.6   Spreadsheets

When working with multiple expressions, Maple provides tools for creating and accessing spreadsheets in your execution groups. Review **?spreadsheet** for a thorough discussion.

2.4.7   Worksheet Management

Worksheet management consists of operations performed on entire worksheets including the workspace. You should consult **?workspace[managing]** for detailed explanations of operations, like opening and closing. Beware that saving work in Maple Worksheet MWS format creates a file only Maple understands. Never print these files!

2.4.8   Miscellaneous

Various appendices in this book demonstrate other features, such as programming and MATLAB, that you might find useful after gaining more experience with Maple.

2.5   LEARNING TOOLS

Sometimes learning new software can be overwhelming—there's so much to learn. With Maple, memorizing the thousands of commands might seem insurmountable. Thankfully, online documentation explains most everything you will need to do. Maple also provides other tools that help you learn. In time, you might drift away from some of these features as you become more experienced. But for now, review the tips in this section and use them in your early Maple explorations.

2.5.1 Palettes

Maple's *palettes* are "clickable" windows that provide templates for choosing common commands that you might wish to enter. If Maple does not show any palette windows, select <u>V</u>iew→<u>P</u>alettes→Show <u>A</u>ll Palettes. Inside the Maple workspace, four palettes will appear: the expression, symbol, vector, and matrix palettes. Figures 2.4 and 2.5 show the symbol and expression palettes. Chapter 9 will demonstrate the vector and matrix palettes. You can find links to online documentation in **?palettes**.

2.5.2 Symbol Palette

Using a palette resembles ordering a meal at a restaurant. First, you identify what you want to "eat," or enter. Then you "order" or select it. Suppose you're "hungry" to enter the symbol ξ, but you can't figure out what it's really called other than "squiggle." Well, to order "squiggle" from your palette,

- Move your cursor to the position at which you wish to insert a symbol in your worksheet.
- Click the symbol in the palette.

Maple's name for the symbol will appear in your worksheet. For instance, try inserting "squiggle" using the palette:

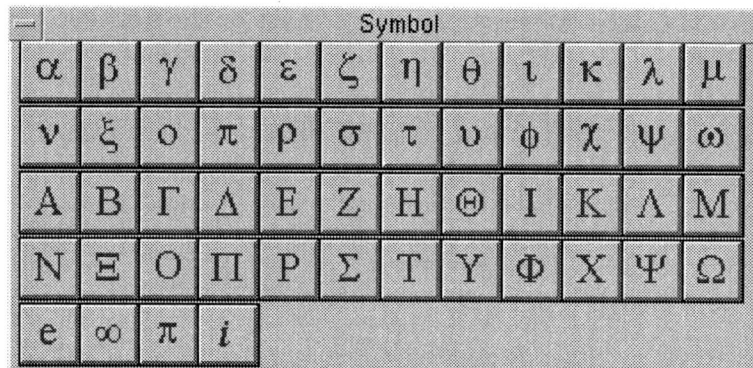

Figure 2.4 Symbol Palette

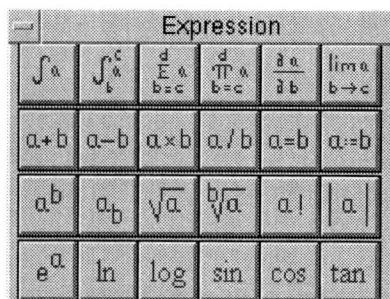

Figure 2.5 Expression Palette

Step 15: Entering a Symbol

Place the cursor in a new input paragraph, find the Symbol Palette, and click on the ξ symbol.

> **xi;**↵   *The Maple input* **xi;** *will appear.*

                                    ξ   *After you press* ↵ *Maple shows the symbol in the correct font.*

Want to discover more names for other symbols? Use the Symbol Palette or check out Appendix A. See also **?sympalette**, **?greek**, and **?symbolfont**.

2.5.3 Expression Palette

What if you want to enter an expression, but you do not know Maple's command names? As you used the symbol palette, you will select items from the expression palette using your mouse. Unlike the symbol palette, however, the expression palette entries are *patterns* that you will fill in at the prompt. For example, clicking the **a+b** pattern will not actually use an **a** and **b**. Rather, Maple inserts placeholders which have the form **%?** for Maple Notation or **?** for Standard Math. For expressions with multiple placeholders, press the Tab key to move to a new placeholder.

For instance, suppose you need to evaluate the expression $4x^{\sqrt{2}}$. Using Maple Notation, perform the following instructions:

Step 16: Entering a Symbol

Select **a** × **b**, *press Tab, select* **a**^{b}, *press Tab, select* $\sqrt{\text{a}}$ *from the Expression Palette.*

> **((%?)*(%?^sqrt(%?)));**   *Maple automatically types this expression as you select*
                                                                                                palette items.

To fill in the values, use the Tab key to jump between placeholders and then type your input:

Step 17: Filling in Placeholders

Move cursor to beginning of input, press Tab, enter **4**, *press Tab, enter* **x**, *press Tab, and enter* **2**.

> **((4)*(x^sqrt(2)));**↵   *You may also manually delete any placeholder (* **%?** *).*

                              $4x^{(\sqrt{2})}$ *Maple generates the output you expected.*

When using Standard Math, the Context Bar will show a box called the *edit field*, which contains the expression in Maple Notation. Watch the edit field as you use the palette to learn Maple commands in Maple Notation. For more information, see **?exppalette**.

2.5.4 Context Sensitive Menu

What should you do when you are looking for a particular function but you do not know the exact name? You could perform a topic or text search, as described in Chapter 1. For an even quicker search,

- Highlight the expression you wish to manipulate.
- Right click the expression.

Maple will display a *context-sensitive menu* which contains selections for "common" commands used with expressions that resemble the one you highlighted. By selecting a menu option, Maple will figure out the input for you and type it in a new execution group. For example, try manipulating the expression $x^2 + 4x + 4$:

Step 18: Context-Sensitive Menu

```
> x*x + 4*x + 4;
```
                                        Trying the expression palette, instead, gives `((%?^%?)+`
                                                              `(((%?)*%?))+%?);`.
$$x^2 + 4x + 4$$
                                                              Everything is OK here.

Select the output $x^2 + 4x + 4$ with the left mouse-button, then right-click the mouse, and select Factor.
```
> (x+2)^2;
```
                                        Maple automatically factors the expression for you.

So, what was the function Maple used? Try entering factor **(x*x + 4*x + 4)** to produce the same result. Although Maple won't tell you the function name, the function name tends to match its equivalent in the context-sensitive menu. For more information, consult Basic Features... Mathematics... context-sensitive menus or **?manipulatecsm**. What if you don't like the choices provided to you? More adventurous students should investigate **?context** and **?examples[context]** to create their own menus.

2.5.5   Smartplots

Some expressions require a lot of tinkering with options to obtain a useful plot. To save time, you may want to let Maple make a few guesses on its own using a ***smartplot***, which is a plot automatically generated by Maple and you let Maple select the options. You may then use context-sensitive menus to manipulate the plot. To generate a smartplot, follow the same steps for getting a context-sensitive menu and select Plots. You should choose a type of plot, like two- or three-dimensional, from a submenu. For instance, you can generate a three-dimensional (3-D) plot as follows:

Step 19: Smartplot

```
> x*cos(y)-sin(y);
```
                        Enter the expression. Using two different variables will make the plot 3-D.
$$x\cos(y) \ - \ \sin(y)$$
                                        You may also use the expression palette.

Select the output, right-click, and select Plots→3-D Plot→x,y.
```
> smartplot3d[x,y](x*cos(y)-sin(y));
```
                                        Maple automatically enters this
                        expression and generates the plot as output in this execution group.
                        Maple indicates that the plot is a smartplot using the word Live.

Live

To alter features of the plot, as shown in Figure 2.6,

- Click the plot—the outer box indicates that the plot may be altered
- Right-click to see a variety of menu options

Practice Plot

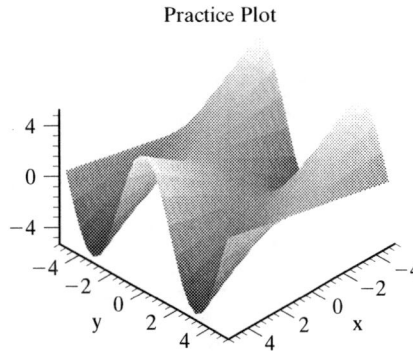

Figure 2.6 Practice Plot

"Normal" plots also have context-sensitive menus. For more information, consult **?contextmenu** and **?smartplot**. Once you learn a bit more about plotting, you should investigate **?plotbuilder**, which provides information on another way to construct plots "from the ground up."

2.5.6 Maple Application Center

Antediluvian[2] versions of Maple provided a *share-library* (**?share**) that contained a collection of worksheets. Now, Maple has combined the entire share-library with an on-line collection of worksheets covering a vast array of applications, which you can find at http://www.mapleapps.com. See Help→Maple *on the Web* for this and other links. You should also see example worksheets that are listed in **?examples[index]**.

PRACTICE!

17. What Maple Input will generate the output $\delta\iota\sigma$?

18. What Maple Input will generate the output $\dfrac{x}{\sqrt{y+2}}$?

19. Generate the plot shown in Figure 2.6.

2.6 APPLICATION: ZEBRA MUSSELS

This section demonstrates a brief example from environmental engineering and science that illustrates key Maple commands.

2.6.1 Background

Sometimes, big engineering problems come in small packages—perhaps even millions of them. Zebra mussels have infested some North American waterways. Consider the image in Figure 2.7. These remarkably sturdy aquatic creatures clog pipes and foul water. Hence, much research is devoted to dealing with these pests.

Assume that the exponential-growth equation

$$P = P_0 \exp(rt) \tag{2.1}$$

governs the zebra-mussel population in a given body of water. Parameter P_0 represents the initial size of the colony, which has a growth rate r over time period t.

---

[2]This literally means "before the Great Flood." Yes, I like using antediluvian terms.

Figure 2.7 Zebra Mussels (courtesy of U.S. Geological Survey, Great Lakes Science Center)

2.6.2 Problem

Given a growth rate $r = 0.9\,(\text{years}^{-1})$, what relative population growth P does the model in Eq. (2.1) predict in a span of 10 years?

2.6.3 Methodology

Inside Maple, retype pertinent parameters and data:

Step 20: Zebra Mussel—Restate

P = population and $P0$ = initial
population (units = critters/m^2)↵   *Enter text.*
r = growth rate (1/year) and t = time (year)↵   *Enter text.*

Next, state your known data, and assign Maple variables to these values:

Step 21: Zebra Mussel—Separate

```
> r:=0.9;↵
```
$$r := 0.9$$

Assign to the name r the value 0.9.
Maple reports your assignment.

Now, assign the model in Eq. (2.1). Using an assignment statement, place the unknown variable P on the left and the function on the right. Express the function e^{rt} as exp(rt):

Step 22: Zebra Mussel—Model

```
> P:=(P0)*(exp(r*t));↵
```
$$P := P0\,\mathbf{e}^{.9t}$$

Assign to the name P the expression P_0exp(rt).
The function exp(x) is written as e^x.
For now, just duplicate the input written here.

Since the problem has not provided an initial zebra mussel count, divide the equation by P_0, and plot P/P_0 for the given time interval $0 \le t \le 10$ years:

Step 23: Zebra Mussel—Solve

```
> plot(P/P0,t=0..10,title="Exponential Growth").⏎
```

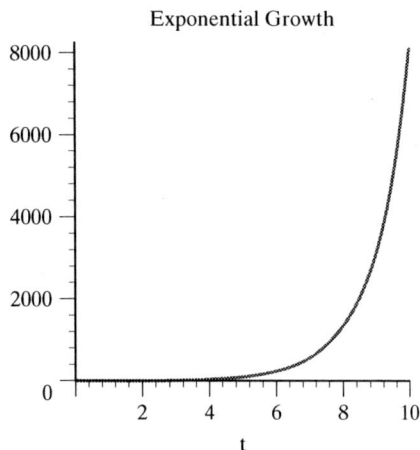

Wow! That looks like a lot of pesky zebra mussels. Finally, find the relative increase at 10 years:

Step 24: Zebra Mussel—Report

```
> t:=10: P/P0;⏎
```
Assign to t the value 10, but suppress the output.
In the same input line, evaluate the relative increase.

$$8103.083928$$
Maple's evaluation of the expression $\dfrac{P}{P_0}$.

2.6.4   Solution

After 10 years, the zebra-mussel population will have grown to about 8,000 times the original population, according to the chosen model. The exponential function models rapid and compounding changes, as that of zebra-mussel reproduction. However, this model provides only a crude approximation. Numerous factors that were not represented in the model contribute to the growth and death of zebra mussels.

KEY TERMS

active input	assignment	case sensitive
context-sensitive menu	expression	execution group
inert input	Maple input	Maple notation
Maple output	palette	paragraph
plot	smartplot	Standard Math
text		

SUMMARY

- The Maple worksheet contains computations and documentation.
- A user may find online help with the help operator **?**.

- The user enters Maple input to evaluate output.
- Execution groups collect input and output.
- Many Maple commands consist of expressions containing variables, functions, and values.
- Assignments store values inside variables.
- Maple provides useful tools, such as palettes and context-sensitive menus, which help users learn the software.
- Maple creates graphs of expressions with plots.

Problems

Write your answers to the problems inside a Maple worksheet. Be neat and beware of case-sensitive variable names. Use the following format, unless instructed otherwise:

> Problem 0.1
> Solve for $1 + 1$.
> `> 1+1;`
>
> 2

1.   Enter the following information on the top of your worksheet:

> HOMEWORK #:
> Name:
> Section:
> Date:

2.   What is Maple?

3.   Specify three uses of Maple.

4.   What is an execution group?

5.   Indicate the menu selections to insert both a section and a subsection.

6.   How do active and inert input differ?

7.   How do Maple input and output differ?

8.   How do Standard Math and Maple Notation differ?

9.   Explain how to find help on the following:
       (a) Maple's help facilities.
       (b) Using the **?** command.

10.   Find help on Maple names.

11.   Suppose that you save your worksheet to a file called **hw1.mws**, using the menu-selection sequence File → Save As.... Assume that you then exit Maple. How can you print the contents of **hw1.mws**?

12.   Assign the value 10 to the name *B1*. After entering the assignment, show the output. Then, enter the expression again and suppress the output.

13.   Assign the expression $1 + \frac{1}{2}$ to the name *B2* and show the output.

14.   Evaluate $1 + 2$, $1 - 2$, 1×2, and $1 \div 2$ using Maple. Show the output for each expression.

15.   Start a fresh Maple session. Enter the following Maple input:
       `> A:=2:`

Generate output from the following input:

```
[ >   A; a;
```

Does Maple report the same values for A and a? Why or why not?

16. Assign $\frac{100}{10}$ to $B3$, but prevent Maple from showing the output. Next, display the value of $B3$ in a new execution group.

17. Find help on the sine function using command-line help. Next, demonstrate that Maple evaluates $\sin\left(\frac{\pi}{2}\right)$ as the value of 1. Hint: Use the symbol palette to enter π.

18. Plot $y = x^2$ for $0 \le x \le 2$. Label your plot "Parabola."

19. Plot $y = mx + b$ for $m = 1$ and $b = -1$ over the domain $-2 \le x \le 2$.

20. Produce execution groups in Maple that look like the following:

(a)
```
[ > restart;
```

(b)
```
[ > 1+1:
  > 'Hello':
  > EQN := y=m*x+b:
  > A:=10:
```

(c) Hello, I am just text.

Now, try **boldface** and *italic* styles.
```
[ > # I am just a happy comment:-)
```

(d) Use the backslash (\) as a continuation character for Maple input:
```
> 1 + 2 +\
> 3 + 4;
```
$$10$$

(e) Here is an embedded inert equation $\frac{a+b}{c}$. Hint: Investigate the Insert menu selections.

Here is the same equation, $\frac{a+b}{c}$, but now it's active input.

Hint: Investigate the icons on the Context Bar.

$$\frac{a+b}{c}$$

(f) Produce this execution group, followed by the section with its subsections, as in the example that follows. Hint: See **?structuring2**.

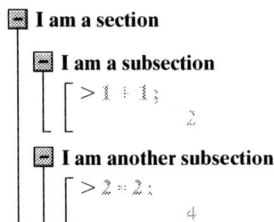

I am a section

    I am a subsection
```
[ > 1 + 1;
```
$$2$$

    I am another subsection
```
[ > 2 * 2;
```
$$4$$

3

Maple Language

3.1   INTRODUCTION

Understanding how Maple interprets commands will greatly assist your work. This section introduces key features of the Maple language that you will use to solve a variety of problems throughout this text and, we hope, in your later studies.

OBJECTIVES

After reading this chapter, you should be able to

- Distinguish elements of the Maple language
- Use Maple operators in accordance with operator precedence and associativity
- Choose Maple names according to variable-naming conventions
- Classify the kinds of Maple statements
- Decompose expressions into expression trees
- Manage variable assignments
- Unassign previously assigned names
- Describe Maple's rules for automatic simplification and full evaluation

3.1.1 Language

Maple commands constitute a written ***language***. A written language provides a means to communicate concepts using combinations of symbols in a defined and universally accepted fashion. For instance, the English language combines words and punctuation to form sentences using rules of grammar. When writers form properly constructed sentences, they hope their readers will extract meaning from them, assuming that the writer is competent.

3.1.2 Maple Language

When speaking of Maple, consider Maple's collections of symbols as words that are formed from standard keyboard characters. Putting these words into commands gives you Maple's sentences, which even include punctuation! For instance, you could think of the input **sqrt(4);** as a way of telling Maple, "Hey, Maple! Find the square root of 4, show the output, and stop." Yes, computer languages tend to compress quite a bit of conversation!

As you would communicate with a person, Maple requires that the user consider the grammar and meaning of their sentences, which are often referred to as ***syntax*** and ***semantics***, respectively:

- Syntax governs the structure of commands and command entry.
- Semantics govern how the program understands and acts upon commands.

For example, refreshing a worksheet involves a one-word "sentence" composed of a known command (**restart**) punctuated by a semicolon (**;**):

Step 25: Restart Maple

> **restart;**   *Refresh the current worksheet by erasing all current assignments.*

As shown in this chapter, you will build expressions using elements of Maple's language.

3.2   LANGUAGE ELEMENTS

Using a set of characters to write Maple commands, the following items make up the written-language elements of Maple: *character set*, *tokens*, and *special characters*.

3.2.1 Characters

Maple's alphabet uses standard keyboard characters to form all other language elements:

- Uppercase letters: **ABCDEFGHIJKLMNOPQRSTUVWXYZ**
- Lowercase letters: **abcdefghijklmnopqrstuvwxyz**
- Digits: **0123456789**
- Miscellaneous characters: **!@#$%^&*()_+|~/=\`{}[]:";'<>?,./space**

If you are unfamiliar with some of the character names, you should review Appendix A. This alphabet provides the characters used to build the language elements, discussed next.

TABLE 3.1 Maple Tokens

Token	Definition	Reference
Operator	Symbol that performs mathematical tasks	`?operator`
Name	Label for functions, mathematical variables, and system variables	`?name`
Reserved Word	Reserved and unassignable command names	`?keyword`
Integer	Whole numbers that use the digits 0, 1, 2, 3, 4, 5, 6, 7, 8, and 9	`?integer`
String	Characters enclosed by double quotes (" ")	`?string`

3.2.2 Tokens

Maple's **tokens** form many of the language's words, which help you build commands, much as you would construct a sentence out of words. Table 3.1 summarizes the kinds of tokens Maple uses. From these elements, you may construct all kinds of important expressions, as demonstrated throughout this text. This chapter provides an overview of some of these tokens.

3.2.3 Special Characters

There are three kinds of special characters: **white space**, **punctuation**, and **escape characters**. These characters affect how Maple processes input and may even cause "exterior" actions to occur. Do not confuse the characters that this section discusses with operators!

White space refers to blank characters and lines that separate tokens.

- Blank characters refer to spaces and tabs. Use the spacebar and Tab keys.
- New lines refer to return and line-feed, which are considered identical. Use Enter or Return (↵).

You may use as many blanks and new lines as you wish, but never break apart an element, like the assignment operator (`:=`)!

Punctuation characters provide another form of token separator. For instance, semicolons (`;`) and colons (`:`) terminate commands. You could actually enter **1+1;2+2;** on the same command line, which causes Maple to add 1 and 1, add 2 and 2, and show the output for both computations. Maple's punctuation characters are summarized in Table 3.2. Note that you will not use many of them until reaching other parts of this text.

TABLE 3.2 Punctuation

Character	Uses	Example	Reference
`;`, `:`	Separate statements	`1+1;1+2:`	`?;`, `?separator`
`,`	Form sequence of expressions	`plot(x,x=1..5)`	`?,`, `?exprseq`
`` ` ``	Delay evaluation	`` `sqrt(4)` ``	`` ?` ``, `?uneval`
`` ` ``	Form names	`` `D I S` := `1!`; ``	`` ?` ``, `?name`
`()`	Separate expressions, function parameters	`(1 + 1)/(sqrt(4);`	`?function`
`[]`	Lists	`[1,3,1]`	`?list`
`{}`	Sets	`{1,2,3}`	`?set`
`<>`, `|`, `,`	Vectors and matrices	`<<1,3>|<2,4>>`	`?|`, `?MVshortcut`

TABLE 3.3 Escape Characters

Character	Uses	Example	Reference
?	Invoke help, find help on topic	**?sqrt**	**??, ?help**
#	Skip comments	**# hello**	**?comment**
\	Continue line, insert control character	**`a+\nb`**	**?\, ?backslash**
!	Execute operating system command	**!ls**	**?system, ?escape**

The ***escape characters*** shown in Table 3.3 resemble tokens, but serve a different purpose. An escape character tells Maple to perform a task "away from" the command line. Maple "leaves" the command line, activates or initiates the required task, and then "returns." For instance, Maple evaluates commands after the prompt until encountering a **#**, which tells Maple to treat all proceeding characters as inert commentary.

PRACTICE!

1. Identify the following items of input in terms of Maple-language elements: **1**, **+**, **TEST**, **sin**, **and**, **:**, **()**, **"Test Plot"**.
2. Does the input **restart; A := 1;** produce an error message?
3. Why does the input **A : = 1** produce an error message?

3.3   OPERATORS

This section focuses on the ***operator*** token. An operator is a mathematical symbol that performs tasks on zero or more expressions, as shown in this section.

3.3.1   Arithmetic

Maple uses the common notation for arithmetic operations shown in Table 3.4. For instance, the expression $1 + 2$ combines the integers 1 and 2 with the addition operator, the plus sign (+). Elements acted upon by operators are called ***operands***. For instance, try an expression that uses addition and division:

Step 26: Arithmetic

> **1 + 1/2;**   *Add fractions together. Remember to press ⏎ from now on!*

$$\frac{3}{2}$$

Maple added 1 to $\frac{1}{2}$, which produces $\frac{3}{2}$.

Consult **?arithop** for more information on arithmetic operations.

TABLE 3.4 Arithmetic Operators

Operation	Symbol	Standard Math	Maple Notation
Exponentiation	^	$2^3 = 8$	**2^3**
Multiplication	*	$2 \times 3 = 6$	**2*3**
Division	/	$2 \div 3 = \frac{2}{3}$	**2/3**
Addition	+	$2 + 3 = 5$	**2+3**
Subtraction	–	$2 - 3 = -1$	**2-3**

3.3.2   Read Me!

Suppose you wish to multiply **a** by **b**. Never forget to specify operators! Never enter **ab**. Never enter **a(b)** or **(b)a** or **(a)(b)**. Never even *think* of entering **a times b**. However, *do* enter **a*b** or **b*a**. As discussed shortly, you can also specify parentheses, as in **(a)*b**, **a*(b)**, **(a)*(b)**, and **(a*b)**.

3.3.3   Functional Syntax

Occasionally, you might see an alternative syntax that Maple allows for operators using a functional form. For instance, rather than entering **1+2**, you may enter **`+`(1,2)** using backquotes (**`**), which are discussed in Section 3.4.4. In general, the functional syntax is **`operator` (operands)**. Why bother? Actually, you usually shouldn't, but the syntax does help you form expression trees, which are discussed in Section 3.7.2.

3.3.4   Miscellaneous Operators

Consult **?index[expression]** or **?operator** for a complete operator list. Consult **?operators[binary]**, **?operators[unary]**, and **?operators[nullary]** for listing according to operator type. Consult **?define** to create custom operators.

PRACTICE!

4. In Maple Notation, determine the operators that the expression $-1 + 2^3$ employs.
5. Assign the value 1 to a. Assign the value 2 to b. Multiply a and b. Now, enter ab. Compare the output for entering **a*b** and **ab**.
6. Enter $4 \leq 5$ and $5 \geq 4$ into Maple. Compare the output from both inputs. Hint: Consult **?inequality**.

3.3.5   Operator Precedence

Beware of ***operator precedence***—rules which dictate that certain operators always act upon expressions *before* other operators do. For example, multiplication and division precede addition and subtraction. If you don't believe me, try incorrectly entering the expression $\frac{1}{2+3}$ such that you ignore the rules of operator precedence:

Step 27: Treat Operators with Care!

```
> 1 / 2+3;
```
$$\frac{7}{2}$$

Attempt to solve the problem $\frac{1}{2+3}$.

Why does Maple not produce $\frac{1}{5}$?

The expression **1/2+3** first evaluates **1/2** and then adds **3** to the result because division precedes addition. Instead, you must surround **2+3** with parentheses to ensure Maple's evaluation of $\frac{1}{5}$:

```
> 1 / (2+3);
```
$$\frac{1}{5}$$

Use parentheses to solve the problem $\frac{1}{2+3}$.

Now you have the correct answer!

You should always surround ambiguous expressions with parentheses, but never use square brackets (**[]**), curly braces (**{}**), or angle brackets (**<>**) when you need parentheses. Consult **?operators[precedence]** and **?syntax** for more information.

3.3.6   Operator Associativity

Operator associativity resolves the treatment of operators that have equivalent precedence. Many operations are *left associative*, like subtraction (**−**), which means that they compute *left* to *right*.

For instance, the expression $1 - 2 - 3$ evaluates to -4 because $1 - 2$ will be calculated before reaching -3. To demonstrate, enter the input without parentheses:

Step 28: Operator Associativity

> **1 - 2 - 3;**   *Subtract three numbers. Which order does Maple pick for the operations?*
                          -4 *Maple automatically evaluates* (1-2)-3 = -4 *by default.*

Next, change the order of Maple's operations by surrounding **2-3** with parentheses:

> **1 - (2-3);**   *Subtract three numbers. Deliberately supply parentheses.*
                          2 *Maple now evaluates* 1-(2-3) = 2.

Some operators, like exponentiation (**^**), require parentheses if the operators are used in conjunction with another operator. For instance, you must enter **2^(-2)** to express 2^{-2}. These kinds of operators are called *nonassociative*.

PRACTICE!

7.   What Maple input will produce the following output:

$$\frac{a^2}{A + \dfrac{1}{a}}?$$

8.   What happens if you enter **sin((a+b));**? What correction should you make?

9.   Will **[1+3]/4** produce the same output as **(1+3)/4**? Why or why not?

10.   Is **1^2^3** acceptable Maple input? If not, correct it. Hint: Consult **?operators[precedence]**.

3.4   NAMES

Equations like $y = mx + b$ use letters as variables that represent quantities that could change, or literally, *vary*. Maple expressions are often constructed from such variables. This section reviews variable naming conventions using the name token.

3.4.1   Symbols

Maple's language defines the most basic form of a variable as a ***symbol***, which follows these rules:

- You may construct symbols using lowercase letters (**a-z**), uppercase letters (**A-Z**), digits (**0-9**), the underscore character (**_**), and question mark (**?**).

- You never use a digit or question mark as the first character or symbol.
- Maple is case sensitive.

Consult **?type[symbol]** for more information.

3.4.2 Name

When speaking symbolic mathematics, a Maple **name** typically serves the role of a variable. A name is a symbol with the additional property of *indexing*, as discussed in Chapter 4.[1] Maple already uses many names to label functions and constants. You will use names to build and assign expressions. Examples of valid names include **A**, **A1**, **A11**, **A1A**, **A_1**, **a**, and **alpha**. As a brief example, enter an unassigned name, which Maple evaluates as just the name you entered:

Step 29: Names

```
> Name;                                          Check the value of name Name.
              Name              Name has no assigned value, so Maple reports the name.
```

Consult **?name** or **?symbol**, **?type[name]**, and **?indexed** for a complete set of rules and more information.

3.4.3 Protected Names

Maple does not allow you to use every possible name. Imagine the potential chaos if you decided that **sqrt** should perform multiplication by 7 instead of taking a square root. To prevent you from making potentially disastrous assignments, Maple predefines certain names, called **protected names**, for library functions, variables, and constants. Maple prevents protected names from assignments, like **sin** and **sqrt**. For a list of what you cannot directly assign, investigate **?ininames** and **?inifcn**. You should also avoid trying to reassign keywords, as discussed in Section 3.5.

Rather than skimming multitudes of Help windows, a new user typically, and often inadvertently, discovers "surprise" protection during a session:

Step 30: Detecting Protected Names

```
> D := 1;                                          Attempt to assign something to D.
Error, attempting to assign
to `D` which is protected                                    Maple reports an error.
```

Otherwise, you may check for protection with **type(name, protected)**:

```
> type(D,protected);              Check if Maple protects D from assignment.
              true                          Maple confirms that D is protected.
```

Also, try entering **?name** to see if Maple already uses **name** for a function or command name.

```
> ?D                                          Maple opens a Help window on D.
```

---

[1]An *indexed name* means a symbol with a subscript, like x_1 and y_w. These are valid names in Maple that are formed using square brackets (**[]**). To form those examples, enter **x[1]** and **y[w]**, respectively.

If you really want to use a protected name for an assignment, investigate **?protect** or **?unprotect**, but I don't recommend using either of them.

3.4.4   Backquotes

To create more descriptive names, surround characters with backquotes, as in **`I am a name`**. You will usually find the backquote (` or ') on the same keyboard key with a tilde ($\sim$). Backquotes help you create names that Maple would normally consider illegal, such as **`1A`**, and **`Blurst blurg wozzie`**. Such names are also available for assignments:

Step 31: Assign Backquote Names

> **`Variable 1` := 10;**   *Assign a value to a backquoted name.*
>   *Variable 1* : = 10   *Maple reports your assignment.*

In general,

- Do not use forward quotes (', ʹ, or ʼ), which are found on the same key with a double quote (", ", or **"**).
- You may backquote some reserved words, but you should avoid doing so.
- For an unprotected name or unreserved word, Maple automatically replaces that name with the actual name written without backquotes.

For further rules, consult **?name**.

3.4.5   Miscellaneous

After you have become familiar with Maple, you will likely encounter many items related to Maple names. These items include local and global variables (**?procedure**), environment variables (**?envvar**), indexed names (**?name**, **?indexed**), aliases (**?alias**), assumptions (**?assume**); and labels (**?labeling**). You will discover these items as you work through this text.

PRACTICE!

> 11.   Which of the following is a valid Maple name, **1** and/or **`1`**? Hint: Enter **type(expr,name)**.
>
> 12.   What input will generate *Success := Practice + Patience* as output?
>
> 13.   Are the Maple names *Ira* and *ira* equivalent? Why or why not?
>
> 14.   Is γ protected? Hint: Refer to the symbol palette.
>
> 15.   What does the input **select(type, {unames(), anames (anything)}, protected)** do? Try it out and then see **?type [protected]**.

3.5   RESERVED WORDS

Maple does not consider reserved words, also called *keywords*, to be protected names. Unlike a protected name, Maple never allows you to change a reserved word's meaning because the word forms an important element of the Maple language. For instance, some reserved words include operators, like **or** and **union**. Many reserved words are used for programming, as demonstrated in Appendix D. Consult **?keyword** for a listing of Maple's reserved words. If you attempt to assign a reserved word, you will receive an error message:

Step 32: Reserved Words

```
> and := 10;                              Attempt to assign reserved word and.
Error, reserved word `and` unexpected    Maple reports an error.
```

As discussed in **?name**, you may backquote a reserved word to give a semblance of an assignable name, but I do not recommend doing so.

PRACTICE!

> 16.   Are **in** and **or** reserved words?
>
> 17.   Can you assign values to **in** and **D**?

3.6   STATEMENTS

Recall the language analogy for a moment. In a written language, a portion of text called a *sentence* usually ends with a punctuation character, as in ending an English sentence with a period. Well, Maple's sentences are called **statements**. A statement combines tokens to form a meaningful portion of input terminated by a semicolon (**;**) or colon (**:**). Continuing the analogy, you might even wish to think of a worksheet as a "book," with a collection of "sentences."

Maple processes each statement in succession, where each statement instructs Maple to perform a task. Overall, Maple classifies several categories of statements, some of which are described briefly in Table 3.5. For a majority of this text, you will concentrate on expression and assignment statements. In later studies, you will need the selection and repetition statements that help you develop programs. Investigate **?index[statement]** for a listing of help topics and commands that deal with statements.

TABLE 3.5   Maple Statements

Statement	Definition	Example	Reference
expression	Maple will evaluate the expression.	> sqrt(4);	?entering
assignment	Maple will evaluate an expression and store the result in a name.	> x := sqrt(4);	?:=
empty	Maple will do nothing and skip to the next statement.	> ;	?;, ?empty
quit	You will exit from Maple in the command-line version, only.	> quit;	?quit, Figure 1.4
selection	Choose a statement based on a condition.	> x := 1: if x <= 1 > then x:=2; end if;	?if, ?try, Appendix D
repetition	Repeat a statement.	> for i from 1 to 3 do > i; end do;	?do, Appendix D
save	Save variable assignments into a file.	> save "work.m"	?save, Appendix D
read	Read in a file into Maple.	> read "work.m"	?read, Appendix D

3.7   EXPRESSIONS

This section elaborates further on the structure and construction of expressions. Later sections rely on your understanding of how Maple interprets expressions, so study this section carefully.

3.7.1   Expression Elements

Recall that you build an *expression* using Maple-language elements called tokens, which are built from Maple's character set. Generally, Maple expressions are built from names, operators, integers, and some punctuation. For instance, consider the mathematical expression, $\sqrt{1 + 1/10}$. The Maple expression **sqrt(1+1/10)** uses the function name **sqrt** along with operators (**+** and **/**) that connect the integers **1** and **10**. Parentheses **()** punctuate this expression. Including a terminator as **sqrt (1+1/10);** creates an *expression statement*, which you may enter for Maple to evaluate. You will discover an even more interesting expression "building-blocks" that Maple provides throughout this text. Consult **?syntax** and **?index[expression]** for more information.

3.7.2   Expression Trees

Figure 3.1 demonstrates how to draw an expression like $\sqrt{1 + 1/10}$ as an ***expression tree***, as shown in Figure 3.1. Turn the book upside-down to see the tree. Following syntax rules, like operator precedence and associativity, Maple *parses*, or "splits," the expression into subexpressions. For this example, some subexpressions include $1 + 1/10$ and $1/10$. Maple continues parsing the subexpressions until reaching tokens. Names, operators, and integers form the nodes on the expression tree,[2] which are 1 and 10 in this example. The branches indicate which elements are acted upon, or connected to, a particular node. Advanced students might wish to investigate **?packages[types]** for related information.

3.7.3   Expression Palette

If you need help entering expressions, refer to **?palette** for more information.

PRACTICE!

18. Draw an expression tree by hand for the expression $\dfrac{1}{a + bc}$.
19. Which of the following Maple inputs **x=10**, **x**, **10**, and **sqrt(10+x/ (10-x))** is an expression? Hint: Consult **?equation** for **x=10**.
20. Why is an assignment like **x := 1** not an expression?
21. What does the input **sqrt('+'(1,'/'(1,10)))** generate? Hint: See Section 3.3.3.

---

[2]More formally, Maple *types* form the nodes of an expression tree. The notion of expression type is explored in other sections.

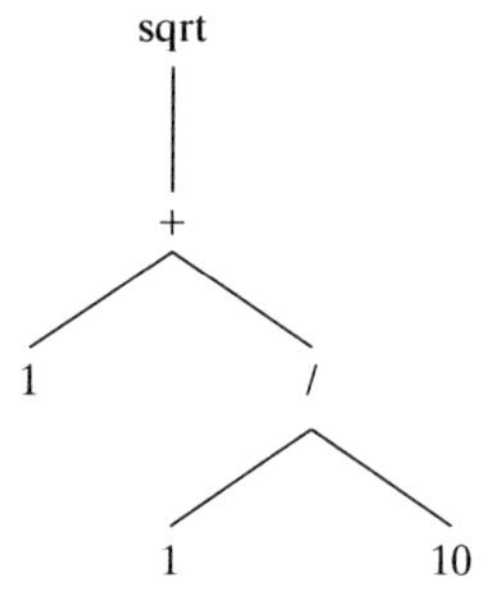

sqrt

+

1   /

1   10

Figure 3.1   Expression Tree for $\sqrt{1 + \frac{1}{10}}$

3.8   ASSIGNMENTS

Expression statements help you perform calculations in an interactive session. However, what if you wish to reuse previous input or split long expressions in a series of smaller inputs? Rather than having you retype, cut, and paste expressions all the time, Maple conveniently provides assignment statements to clarify and ease your work. This section demonstrates the rules and tips for assigning expressions. Students interested in using assignments for programming should review Appendix D.

3.8.1   Assignment Syntax

Maple's general assignment statement **name:=expression** evaluates **expression** and stores the result inside **name**. After entering an assignment, Maple replaces each instance of **name** with the results of **expression**. For example, store the value 1 in the name A with **A := 1**:

Step 33: Assignment

```
> A := 1;
```
$$A := 1$$

Assign the value 1 to the name A.
Maple evaluates your assignment.

When thinking or saying "**name:=expr**," try to say, "**name** gets **expr**." Sometimes you will see "Assign **expr** to **name**" or "Assign to **name** the expression **expr**." Hopefully this terminology will prevent you from entering assignments in the reverse and illegal order, e.g., **1:=A**! When used for assignment, a Maple name is sometimes called a **variable** because the name represents a quantity that can change in value. Maple provides different kinds of variables, as discussed in Appendix D. Consult **?:=** and **?assign** for more rules of syntax.

3.8.2   Assigning Expressions

The expression that you assign does not have to be numerical. You may assign any expression, including other names:

Step 34: Assigning Other Names

```
> Name1 := Name2;
```
$$Name1 := Name2$$

*Assign to the name **Name1** the "value" **Name2**.*
Maple evaluates your assignment.

Eventually, *Name2* might obtain a value, which will change the value of *Name1*, as discussed in Section 3.11. For now, imagine the possibilities of assigning more complicated and exotic expressions, as in **y:=m*x+b**, **z:=x^2/sqrt(y)**, and even **p:=plot (exp(x),x=1..2)**!

3.8.3 Using Assignments

After assigning a name, Maple replaces all future occurrences of the name with the assigned value. Therefore, you may enter **name** as an expression statement to check the value assigned to **name**:

Step 35: Check Assignment

> **Var := 1;**   **Var** *gets* 1.

$$Var := 1$$ *Maple reports the evaluation of your assignment.*

> **Var;**   *Check the value assigned to* **Var**.

$$1$$ *Maple evaluates* **Var**, *which produces the value* 1.

By entering **Var** as an expression statement, Maple evaluated the expression represented by **Var**. You will discover more about how Maple performs that evaluation in Section 3.11. Also, until you reassign or remove the value of *Var*, Maple uses *only* the assigned value 1 in place of *Var*:

Step 36: Maple Remembers Assignments!

> **Var+2;**   *Form an expression using* **Var**.

$$3$$ *Maple replaced* **Var** *with the value* 1.

How long will Maple remember your assignment? Until you enter **restart** or quit Maple, as discussed in Section 3.10. You should also consult **?assigned**, **?anames**, and **?unames** for checking assignments.

3.8.4 Assign and Equal are Different!

Expressions built with the equals sign (**=**) are called *equations* in Maple, not assignments! Equations are a specific type of expression, as discussed in later sections. What happens if you enter equals (**=**) instead of the assign operator (**:=**)?

Step 37: Equations are not Assignments!

> **test = 1;**   *The input* **test=1** *is an expression, not an assignment.*

$$test = 1$$ *Maple evaluates the expression.*

It may look like Maple assigned *test*, but it didn't. Check the value of *test* as proof:

> **test;**   *Check the value of* **test**.

$$test$$ *Maple did NOT assign* **test**.

If you wanted to assign *test*, you should have entered **test:=1** instead. Check **?equation** to learn more about the proper use of **=**.

3.8.5   Unassigning Variables

Once assigned, a name remains assigned until the name gains another value, or the name is cleared. ***Unassigning*** removes any previously assigned expression stored in a name. Unassign **name** by surrounding **name** with forward quotes, as in **name :=** **'name'**:

Step 38: Unassign a Name

> `Var := 1;`   Assign the value 1 to the name **Var**.
$$Var := 1$$ Maple evaluates the assignment.

> `Var := 'Var';`   Unassign the name **Var**.
$$Var := Var$$ Unassigning **Var** removed the stored expression.

> `Var;`   Is **Var** still assigned to any value?
$$Var$$ No: Unassigning **Var** cleared its value.

You might wonder why entering **Var:=Var** will not unassign **Var**. When you review Section 3.11.7, you will discover why the forward quotes are necessary. You should also consult **?unassign** and **?evaln** for alternative approaches to unassigning names.

PRACTICE!

22. Explain the difference between the input statements **J=72** and **J:=72**.
23. All in one statement, evaluate the expression $1 + \sqrt{4b}$ and assign the result to the name *dis*.
24. Store the expression $a + b$ in c. Assign to b the value 1. Then, store the value of $c + 1$ in d. What expression does d now store? Why?
25. Check the values of a, b, c, and d.
26. Unassign a, b, c, and d without using **restart**.

3.9   MAPLE QUOTES

By now, you have seen a variety of Maple quotes which, I admit, can be quite confusing. Table 3.6 summarizes the quotes, how they may appear in Maple windows, and their uses. Refer also to **?quotes** for online help.

TABLE 3.6   Maple Quotes

Quote	Symbols	Use	Example	Reference
Back	`, `	Form a symbol	`` `Hi Jenn!` ``	
		Surround commands	`` `+`(1,2) ``	?`
Forward	', ', '	Unassign	`x := 'x'`	
		Delay evaluation	`'x'`	?'
Double	", ", "	Form a string	`"hello"`	?"

3.10 WORKSHEET MANAGEMENT

Assigned names may conflict with other assignments on different worksheets during the same Maple session. This section presents methods for efficient worksheet management.

3.10.1 Restart

Entering **restart** or pressing the icon ▦ resets a Maple worksheet and removes all assignments and accessed library packages. Appendix B and later chapters demonstrate the library packages.

Step 39: Restarting a Worksheet

> **Var := 1;**   *Assign 1 to the name* **Var**.
                    $Var := 1$ *Maple evaluates the assignment.*

> **restart;**   *Restart the worksheet. All assignments are erased.*

> **Var;**   *Check the value of* **Var**.
                    Var *All assignments were erased.*

For more information, consult **?restart**.

3.10.2 Kernel Modes

During a session, opening a new worksheet does not necessarily start a new session with no variable assignments. Three kernel modes determine how worksheets share name assignments:

- *Shared-Kernel Mode*:   Worksheets share *all* assignments, regardless of where the statements are entered. Entering **restart** removes all assignments from all open worksheets.
- *Parallel-Kernel Mode*:   Worksheets operate independently and do not share assignments from other worksheets. Entering **restart** removes all assignments *only* in the worksheet in which **restart** was entered.
- *Mixed-Kernel Mode*:   Worksheets use either shared- or parallel-kernel mode.

Usually, Maple starts in shared-kernel mode. Consult **?kernelmodes** and **?configuring** for more details on choosing another mode.

3.10.3 Dittos

To avoid retyping many expressions, use *dittos*, which are *nullary* operators: **%**, **%%**, **%%%**. A single ditto reissues the previously entered expression:

Step 40: Dittos

> **a+b;**   *Evaluate the expression* a+b.
                    $a + b$

> **%;**   *Reissue the previously evaluated expression.*
                    $a + b$ *Maple evaluates* a+b *again.*

I suggest avoiding dittos until you have gained more experience managing your assignments. Consult **?ditto** and **?operators[nullary]** for more information.

3.10.4 Assigned Names

To tell Maple to report that all variables have been assigned during the current session, enter **anames()**. For example, in a fresh worksheet, assign some variables and then ask Maple to tell you the currently assigned variables:

Step 41: Assigned Names

```
> restart;                                    Refresh your worksheet.

> a:=1: b:=2:                                 Assign variables a and b.

> anames();                                   Ask Maple to tell you the assigned names.
                        a, b                   Maple reports the assigned names.
```

For information and options, consult **?anames**. You will find related information with **?unames** and **?assigned**.

PRACTICE!

> 27. Determine your system's default kernel mode.
> 28. Run Maple in parallel-kernel mode. Open two worksheets. Enter the statement **A:=1** in one worksheet. Predict the value of **A** in the other worksheet before checking.
> 29. What output would you expect from the following Maple input statements?
>      ```
>      > A:=3: B:=2: C:=1: %%%;
>      ```

3.11   EVALUATING EXPRESSIONS

To use Maple's full power, you need to learn how Maple interprets your input. Understanding how the evaluation of that input occurs will relieve you of any possible confusion. Until now, you have seen Maple evaluating basic input using straightforward commands. But, what if you saw input that looked like the following?

```
> a:=1: x:=a+y: y:=1: a:=2: x;
```

You might wonder if the output is 2 or 3 . . . ? You should check Maple to discover that the correct output is indeed 2. If you didn't anticipate that result, don't worry—we have not yet learned how Maple evaluates expressions. With quite a bit of verbosity, this section tries to explain how Maple maintains exact, symbolic expressions whenever possible. Toward this goal, Maple uses built-in rules to compute output: *automatic simplification* and *full evaluation*. Learning how these rules work will grant you the power to amaze your friends and instructors with your ability to predict and control Maple.

3.11.1 Automatic Simplification

Usually, someone performing calculations wants the simplest, or "smallest" possible results. For instance, 1/2 is simpler than 2/4. Simplification attempts to reduce expressions to smaller forms and might require ingenuity. Maple thinks like a human being in this respect and employs *automatic simplification* rules for reducing "easy" expressions. Maple applies these rules when expression elements are "clearly" identical, as in many arithmetic operations:

Step 42: Automatic Simplification

```
> x:='x':                                        Ensure x is unassigned.
> x+x, x-x, x*x, x/x;                       Enter a sequence of expressions.
                 2x, 0, x², 1         Maple automatically simplified each expression.
```

Automatic simplification also activates for "obvious" expressions, like $\sin(0)$ and $\cos(0)$. For other expressions, knowing exactly when Maple applies automatic simplification is difficult to predict, which is why "easy," "clearly," and "obvious" are shown in quotes. Usually, you just have to enter your input and inspect the output, which actually can be fun, believe it or not. Consult **?assume** and **?simplify** for information on functions that enhance Maple's simplification capability.

PRACTICE!

30. Does Maple automatically simplify expressions, such as $\sin(x + x)$?

31. Enter $\frac{2(x + y)}{4(x + y)}$ such that Maple performs automatic simplification.

32. Does Maple simplify $\sqrt{x^2}$ to x? Why or why not? How might you use **assume** to generate x from $\sqrt{x^2}$?

3.11.2 Full Evaluation

When you enter a statement, Maple performs *full evaluation* on all expressions in the input. Full evaluation is a process that does the following:

- It converts expressions into trees.
- It replaces all assigned names with their respective expressions, which are also converted into trees until all possible replacements have been made.
- It attempts to perform all possible automatic simplifications.
- It performs all mathematical operations for each subexpression.

To *fully evaluate* an expression, Maple usually starts from the bottom of the entire tree and works up until reaching the top of the tree. Maple will perform all operations and functions on all subexpressions to calculate the final result.

For example, suppose you entered the expression **sqrt(1+1/10)**. Maple would convert the expression into a tree, as shown in Figure 3.1. To fully evaluate the input, Maple performs these operations in succession: divide 1 by 10, add the result to 1, then take the square root of the entire result.

3.11.3 Assignments

How does Maple treat assignments when fully evaluating an expression? Consider these two Maple statements:

- Expression statement: **expr**. Maple fully evaluates **expr**, but performs no assignment.
- Assignment statement: **name := expr**. Maple fully evaluates **expr** and stores the results in **name**, if **name** is assignable.

But, what happens when Maple encounters a name *inside* **expr**? For example, you might enter **x:=y+z**. Maple needs to check whether or not the names y and z have assigned expressions to evaluate $y + z$ before storing the final result in x. In the expressions assigned to y and/or z, those expressions might have further assignments, and so on. In general, there are two actions that Maple performs on every name inside all expressions and subexpressions:

- If the name is assigned, Maple replaces the name with the assigned expression retrieved from memory. Maple evaluates **expr** with the substituted expression tree.

- If the name is unassigned, the name remains a symbolic value that Maple includes in the evaluation of **expr**.

At no point, however, does Maple *change* any assignment for a name that belongs to the expression tree. Continuing the example, suppose you had entered the following input:

```
> restart: x:=y+z: y:=1: x;
```

In order of input statements, Maple first erases all assignments and assigns the expression $y + z$ to x. Since y and z have no assigned values, Maple stores $y + z$ in memory for the name x. Then, the value 1 is assigned to y, but Maple does not evaluate x yet. *Entering* **x** causes Maple to fully evaluate x by retrieving the expression $y + z$ from memory, replacing y with 1, adding 1 to z, and reporting the evaluation $1 + z$ as output.

There are more nuances: What if you change the assignment for y *after* entering **y:=1**? What if y were assigned *before* entering **x:=y+z**? Or, what if you change y and re-evaluate x? Worse yet, what if all of this is giving you a headache? Don't panic! The next two sections demonstrate these implications of assignments and will answer all of these questions.

3.11.4 Example 1: Assigned Names

This example demonstrates what happens when Maple encounters an assigned name. Consider the example of a sequence of Maple inputs and outputs shown in Table 3.7. The three columns on the right indicate how Maple stores each expression in memory after the input statements are entered. Now, trace the session:

- Input ① restores variables to just their names.
- Input ② assigns 2 to m.
- Input ③ assigns $2x$ to y, because Maple replaced m with 2 during the evaluation of mx in memory.
- Input ④ changes m. But, y cannot change because y has $2x$ stored in memory.
- Input ⑤ resets m. Again, y does not change unless it gets another assignment.

TABLE 3.7   Full Evaluation (Assigned Names)

Order	Input Statements	Output	m	x	y
①	> restart: y;	y	m	x	y
②	> m:=2;	$m := 2$	2	x	y
③	> y:=m*x;	$y := 2x$	2	x	$2x$
④	> m:=3: y;	$2x$	3	x	$2x$
⑤	> m:='m': y;	$2x$	m	x	$2x$

Do you see the implications of assigning a variable before using it in another expression, like *m* in Table 3.7? During full evaluation, Maple replaces all assigned names with their associated expressions. So, Maple "forgets" that you used the names previously. I tend to find this approach "brittle," as opposed to using unassigned names initially, as discussed in the next section.

3.11.5 Example 2: Unassigned Names

This example demonstrates what happens when Maple evaluates an expression that contains an unassigned name. This order of assignments tends to give you more flexible expressions than using previously assigned names. Consider the example of a sequence of Maple inputs and outputs shown in Table 3.8. Now, trace the session:

- Input ① restores variables to just their names.
- Input ② assigns to *y* the expression *mx*.
- Input ③ gives *m* a value of 2. During evaluation of *y*, Maple replaces *m* with 2 and reports the result 2*x* because of full evaluation. But, Maple does not assign the result of 2*x* to *y*, which remains as *mx* in memory.
- Input ④ gives *m* a new value of 3 which Maple uses during evaluation of *y*. But, *y* remains assigned to *mx* in memory.
- Input ⑤ unassigns *m*. Again, the expression for *y* remains *mx*.

Maple fully evaluates new values of *y* for inputs ② through ⑤, but the assignment for *y* never changes after input ②. Why? Inside Maple's memory, the expression *mx* remains assigned to the name *y* throughout future inputs until you assign another expression, unassign *y*, or restart your worksheet. In this fashion, Maple mirrors the human ability to "remember" an expression. This capability allows you to test different values in a previously assigned expression without having to reassign that expression.

TABLE 3.8   Full Evaluation (Unassigned Names)

Order	Input Statements	Output	m	x	y
①	`> restart: y;`	y	m	x	y
②	`> y:=m*x;`	$y := mx$	m	x	mx
③	`> m:=2: y;`	$2x$	2	x	mx
④	`> m:=3: y;`	$3x$	3	x	mx
⑤	`> m:='m': y;`	mx	m	x	mx

PRACTICE!

Predict the output from the following input statements and then test your prediction with Maple:

33. `[> restart: A:=1: B; B:=A: D; A:=2: B;`
34. `[> restart: B:=A: B; A:=1: B; A:=2: B;`
35. `[> restart: x:=y: y:=x: y; x;`

3.11.6 Execution Groups

Maple evaluates input inside an execution group from left to right, and then, from top to bottom. So, the top input line is evaluated from left to right. The next input line is evaluated from left to right, and so forth. Note that pressing ↵ anywhere inside an execution group enters the *entire* group as input, starting at the top, left-most portion of the group.

3.11.7 Unevaluation

To delay full evaluation, surround an expression with forward quotes, such as **'expr'**. What does that mean? Delaying evaluation is called **unevaluation**, which is a process that involves two steps after you enter **'expr'**:

- Maple strips the input **'expr'** of the forward quotes (**' '**).
- Maple attempts automatic simplification on **expr**.

No further evaluation occurs on **expr**! For instance, try delaying the evaluation of **sqrt(4)**:

Step 43: Unevaluation

```
> sqrt(4);
```
$$2$$
First, evaluate $\sqrt{4}$ *without delay!*
Without quotes, Maple fully evaluates the expression.

```
> expr := 'sqrt(4)';
```
$$expr := \sqrt{4}$$
Now, delay the evaluation of $\sqrt{4}$.
Maple stripped **sqrt(4)** *of its quotes.*

```
> expr;
```
$$2$$
Now, fully evaluate expr.
Without quotes, Maple fully evaluates expr.

Using unevaluation, you can enter **name:='name'** to unassign a name, as shown in Section 3.8.5. Maple first strips **'name'** of the forward quotes, which leaves **name:=name** as input. Hence, **name** is assigned to itself using the literal symbol for **name** without any value. Consult **?uneval** and **?eval** for more information.

PRACTICE!

36. Unassign the variable **x** without using **restart**.
37. Does Maple simplify the input **'1+2'**? Hint: Consult **?uneval**.

PROFESSIONAL SUCCESS: MANAGING ASSIGNMENTS

Have you found how Maple "remembers" assignments a bit confusing? Worksheets show only input and output without an indication of *when* computations were entered. So, you need to keep track of statement order in terms of both place and time. The following example demonstrates the danger of forgetting statement order.

First, initiate a small Maple session and create four empty execution groups. Next, *type* these input lines without entering the statements:

```
> A := 1:   #1
> A := 'A':#2
```

```
> B := A:   #3
> B;        #4
```

What input order will assign the value 1 to *B*? Move your cursor up and down with the arrow keys or mouse, and enter statements in the order #2, #3, #1, and #4. Presto! You get *B* := 1.

Such situations might arise when you forget your statement entry order. Since Maple stores the most recent assignment, unexpected results often arise when one does not keep track of this order. If you find that Maple starts reporting bizarre results, check for missing assignments or unassignments. If you still lose track of assignments, try these tips:

- Select Edit→Execute→Worksheet to enter all input statements from top to bottom.
- Enter **anames()** to see a list of all currently assigned variables.
- Frequently check the current values of your names.

- Unassign names before assigning new expressions.
- Look for case-sensitive names and missing operators.
- Use the parallel-kernel mode.
- Delete everything, and enter **restart**—only as a last result, of course!

3.12 APPLICATION: ENGINEERING ECONOMICS

This section demonstrates how Maple and its language elements assist solving an engineering-economics problem.

3.12.1 Background

Engineers often must choose alternative plans based on economic decisions. In this section, you calculate an item's *annual worth* (AW), which is a "leveled" annual payment that represents the item's yearly income and cost. AW converts single payments, like P the present purchase price (P) and future salvage value (SV), into an equivalent annual amount (A). Assuming a yearly interest rate i over an item's life cycle n, calculate annual worth as

$$AW = (P)(A/P, i, n) + (SV)(A/F, i, n) + A, \tag{3.1}$$

where

$$(A/P, i, n) = \frac{(i)(1 + i)^n}{(1 + i)^n - 1} \tag{3.2}$$

and

$$(A/F, i, n) = \frac{i}{(1 + i)^n - 1} \tag{3.3}$$

- The factor $(A/P, i, n)$ converts a present value P into an annual cash flow A.
- The factor $(A/F, i, n)$ converts a future value F into an annual cash flow A.
- The lone A in Eq. (3.1) indicates annual operating costs that do not need conversion.

3.12.2 Problem

Determine the annual worth of a machine that has a purchase price of $72,000, an annual maintenance and labor cost of $1,000, and a salvage value of $14,000, given an interest rate 12% over a 10-year life cycle. Calculate the percent difference of annual worth for a purchase price of $65,000. Percent difference is defined as

$$percent\ difference = \frac{|original - changed|}{original} \times 100 \tag{3.4}$$

3.12.3 Methodology

First, distinguish the cash-flow values between negative cost and positive income. Use an execution group to label important variables and parameters as text:

Step 44: Economics—Restate

```
> restart:
```
Refresh your worksheet.

Outflow: P (initial price) $= -72000$, A (annual cost) $= -1000$
Inflow: SV (salvage value) $= +14000$
Parameters: i (interest rate) $= 12\%$, n (life cycle) $= 10$ years

Before calculating specific quantities, assign the formulas from Eqs. (3.1) and (3.2) in symbolic form. Because of Maple's full evaluation rules, you will be able to assign different parameters without changing how Maple stored the formulas:

Step 45: Economics—Model

```
> AgivenP:=(i*(1+i)^n)/((1+i)^n-1):
```
Assign (A/P, i,n).
```
> AgivenF:=(i)/((1+i)^n-1):
```
Assign (A/F,i,n).
```
> 'Annual Worth' := (P)*AgivenP +
(SV)*AgivenF + A;
```
Assign Eq. 3.1.

$$Annual\ Worth: = \frac{Pi(1 + i)^n}{(1 + i)^n - 1} + \frac{SVi}{(1 + i)^n - 1} + A$$

Check your formula!

Now, assign the cash-flow values:

Step 46: Economics—Separate

```
> P:=-72000: A:=-1000: SV:=14000:
```
Assign cash flows.
```
> i:=0.12: n:=10:
```
Enter **i** *as a decimal percent.*

You can now evaluate the annual worth simply by entering the variable name. Full evaluation will take care of variable substitution because you made your assignments after assigning the formula for annual worth.

Step 47: Economics—Solve and Report

```
> AW1:='Annual Worth';
```
Evaluate the annual worth.
$$AW1 := -13663.08199$$
Store the result in **AW1**.

To produce only a certain amount of decimal places, enter **evalf(*expr,d*)**, where **d** is the number of digits you want Maple to use in a calculation:

```
> evalf(%,7);
```
Evaluate the previous expression with 7 digits.
$$-13663.08$$
You could also enter **evalf(AW1,7)**.

To compute the percent difference for a different purchase price, assign P a new value and reenter **'Annual Worth'**:

> ```
> A:=-65000: AW2:='Annual Worth';
>              AW2:=-12424.19284
> ```
        Evaluate the annual worth.
        Store the result in **AW2**.

Use **abs(*expr*)** to take the absolute value of ***expr***:

> ```
> pdiff := abs( (AW1-AW2)/AW1)*100 ):
> ```
    Percent difference $= \dfrac{|old - new|}{old} \times 100.$

> ```
> evalf(pdiff,4);
> ```
    Round the result.

         9.078
    The answer is about 9%.

3.12.4 Solution

The annual worth of the machine is $-13,663.08. The percent difference in annual worth from lowering the purchase price to $65,000 is about 9%. Maple can actually assist you with these sorts of calculations with built-in functions—check out **?Units[currency]** along with Appendix C and **?finance**.

KEY TERMS

automatic simplification	dittos	escape characters
expression	expression tree	full evaluation
integer	language	mixed-kernel mode
name	operand	operator
operator associativity	operator precedence	parallel-kernel mode
protected names	punctuation	reserved word
semantics	shared-kernel mode	statements
string	symbol	syntax
tokens	unassigning	unevaluation
variable	white space	

SUMMARY

- The Maple language builds expressions with tokens, which include names, operators, reserved words, integers, and strings.
- Operator precedence and associativity determine the order that Maple chooses operators when evaluating expressions.
- Maple names act as variables to which you may assign expressions.
- Reserved words and protected names are tokens that Maple has already defined and should be avoided for assignments.
- Commands for Maple to act upon, like expressions and assignments, are called statements.
- Maple converts an expression into an expression tree that divides the expression into individual tokens.
- An assignment statement evaluates an expression and stores the result in an assignable name for later use.
- Maple evaluates expressions using full evaluation and automatic simplification.
- Automatic simplification is Maple's built-in method of calculating "obvious" expressions.
- Full evaluation decomposes an expression into a tree and evaluates each subexpression.

- Full evaluation leaves unassigned names as symbolic values and replaces assigned names with their respective expressions.
- To unassign a name, enter **name := 'name'**, which works because of unevaluation due to the forward quotes.

Problems

1. Name four tokens of the Maple command language.
2. Explain the difference between a mathematical expression and a Maple expression that you enter as a statement.
3. How do Maple names and symbols differ?
4. How do protected names and reserved words differ?
5. Give an example of an expression that performs addition and division with integers.
6. Assume you discover the expression $[x(y + z)]$ inside another textbook. Should you avoid square brackets when entering this expression in Maple Notation? Why or why not? Hint: See **?list**.
7. Produce the Maple output 'I am a name that includes backquotes' using a Maple name as input. Hint: Consult **?name**.
8. Draw an expression tree for the expression $\sin(a + b)$. You may do this by hand.
9. Explain the difference between shared- and parallel-kernel modes.
10. Give Maple input that will produce the following output:

    (a) A

    (b) $A + 1$

    (c) $A + \dfrac{1}{a}$

    (d) A^2

    (e) $A^3 + \dfrac{1}{2}$

    (f) $\dfrac{x + y}{z}$

    (g) $x + \dfrac{y}{z}$

    (h) $\dfrac{x}{z} + y$

    (i) $A := x + 1$

    (j) $A \leq B$

    (k) $A = B$

    (l) $\sqrt{\sin(x)}$

11. Evaluate the expressions that follow using Maple. Hints: Be careful with operator precedence and associativity. Do not worry if Maple produces a fractional result.

    (a) $1 + 2 + 3$

    (b) $\dfrac{1}{2} + 3$

    (c) $1 - 2 - 3$

    (d) $1 - (2 - 3)$

    (e) $1 \times 2 \times 3$

    (f) $2 \div 3$

    (g) $(\frac{1}{2})(-\frac{3}{4})$

    (h) $2(\frac{3}{1 + 4})$

    (i) 2^3

    (j) $2^{(-1)}$

    (k) $2^{(2-4)}$

    (l) $2^{\frac{-1}{5+\frac{4}{3}}}$

12. Can you assign the value 1 to the name A with the input statement **1:=A**? Why or why not? Demonstrate with Maple.

13. Perform the following tasks:
    (a) Assign the value of 1 to the name *A*.
    (b) Assign the value of 2 to the name *a*.
    (c) Check the values of both names *A* and *a*.
    (d) Unassign the name *A*.
    (e) Unassign the name *a*.

14. Enter the following input statements:

    ```
    > restart: y=mx+b: m=1: b=2: x=3: y;
    ```

    Why does Maple *not* produce the output of 5? Correct the input statements so that Maple does. Show the corrected input with its corresponding output.

15. Assign the value 10 to *a*. Next, store the value of *a* in *b*. Now, change the value of *a* to 20. Is the current value of *b* 10 or 20? Explain and demonstrate your answer with Maple.

16. Assign the value 123 to *Var1*. Next, assign the same value to *Var2* with **Var2:=Var1**. Are the values of *Var1* and *Var2* the same? Why or why not? Demonstrate your results and conclusions with Maple.

17. Fill in the missing information inside Table 3.9. You may write your answers by hand or as text in an execution group. Hint: Refer to Tables 3.7 and 3.8.

18. Ceramics mix metallic and nonmetallic compounds and have found wide-ranging use in many branches of technology. A type of ceramic called *spinel* has the chemical compound $MgAl_2O_4$. Express the spinel compound using Maple names for each individual element. Hints: Use Maple input. For names with an index, use square brackets, as discussed in **?name**. For instance, enter **B[1]** to express B_1. For more fun with chemical elements, see **?ScientificConstants** and Appendix C.

19. If you invert the factor $(A/F, i, n)$ will you determine the factor

    $$(F/A, i\%, n) = \frac{(1 + i)^n - 1}{i}?$$

    Why or why not? Demonstrate with Maple.

20. Many ingredients, such as cement, water, sand, and coarse aggregate (rocks), compose standard structural concrete. Assuming the proper vibration to remove entrapped air, you can approximate the total volume of concrete for a pour using the following densities: cement ρ_c = 195 pcf (pound-mass per cubic foot), water ρ_w = 62.4 pcf, and aggregate (includes rocks and sand) ρ_a = 165 pcf. Suppose that you can mix a batch of concrete that contains 250 lbm (pound-mass) of gravel, 150 lbm of sand, 100 lbm of cement, and 50 lbm of water. How many batches of concrete do you need to fill a rectangular wall with dimensions 20 ft × 5 ft × 1 ft? Assume no volume loss due to rebar reinforcement. Hint:

TABLE 3.9   Full Evaluation (Problem 17)

Order	Input Statements	Output	x	y
①	> x:='x': y:='y':			
②	> x:=10;			
③	> y:=x+5;			
④	> x:=20: y;			
⑤	> x:='x': y;			

Both gravel and sand make up aggregate, so add their weights together. See Appendix C and **?Units** for help on using units in your calculations.
(a) Restart Maple.
(b) Assign the densities. For instance, **rho[w]:=62.4*lbm/ft^3**.
(c) Calculate the volume of each component: Volume = Mass (lbm)/ Density (lbm/ft³).
(d) Calculate the total volume in one batch. Hint: Sum the volumes in the previous step.
(e) Calculate the required volume of concrete (i.e., the volume of the wall).
(f) Calculate the number of batches: Batches = Volume of wall (ft³) ÷ Volume of Batch (ft³/batch).

21.   Assume an elastic bar is axially loaded with a force P, as shown in Figure P3.21.

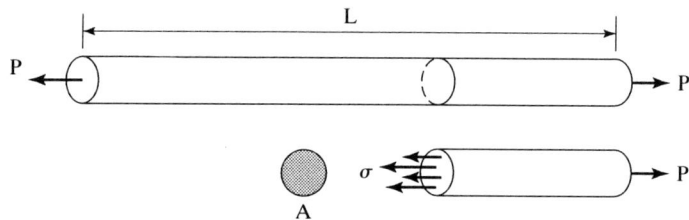

Figure P3.21   Elastic Axial Bar

Engineering *strain* ε is defined as a body's change of length Δ divided by the original length L. The engineering stress σ on the bar uniformly divides the load P by the original area A before deformation. Thus,

$$\varepsilon = \frac{\Delta}{L} \text{ and } \sigma = \frac{P}{A}. \tag{3.5}$$

Assuming a linear relationship $\sigma = E\varepsilon$ between stress and strain, solve these problems:
(a) Assign to the stress σ the expression $E\varepsilon$. E is called Young's Modulus. Hint: Refer to Appendix A and the symbol palette for entering Greek names.
(b) Assuming fundamental units of force and length, what units do stress, strain, and Young's Modulus have? Hint: Consider the formulas for ε and σ.
(c) Plot stress versus strain for Young's Modulus $E = 10$ for $0 \leq \varepsilon \leq 1$.
(d) Substitute the expression for stress P/A and that for strain Δ/L into the relationship $\sigma = E\varepsilon$. You can do this by hand, but type the results as text in Maple. You should determine that $P/A = E\Delta/L$.
(e) Assign the entire expression determined in the preceding problem to the name *EQN*. Your assignment should have the form **EQN := expr1=expr2**. Hints: Include the equal sign (=)! The expression **expr1=expr2** is called an *equation*.
(f) Solve for P using Maple. Hints: Investigate **?solve**. Your input should have the form **solve(EQN, something)**. Your output should be $E\Delta A/L$.
(g) Hooke's Law for a spring states that $P = K\Delta$. Relate this equation to your results in the preceding step. You should be able to show a formula for K in terms of E, A, and L.

4

Expression Types

4.1   TYPES

Basic tokens, like names and operators, help to build expressions. As you solve more complicated models, your expressions will combine these and other tokens in different ways to produce a wider variety of expressions. Maple classifies all expressions according to their mathematical structure. This chapter provides an overview of this classification scheme to help you build expressions that commonly arise in solving science and engineering problems. For on-line overviews, you should investigate Programming...Data Types....

SECTIONS

OBJECTIVES

After reading this chapter, you should be able to

- Classify Maple's expression types
- Identify an expression's surface and/or structured type
- Build expressions using different types of numbers
- Build expressions using Boolean values and relations
- Build expressions consisting of collections of other expressions
- Label and extract elements from lists, sets, sequences, and indexed names

4.1.1   Type Definition

Maple defines a classification of an expression as a *type*. According to `?type [definition]`, a type is a Maple expression recognized by the **type** function. (Yes, I know that sounds like circular reasoning.) For now, think of a Maple type as a common classification to which an expression may belong. Every Maple expression has at least one type, and sometimes, many more. For example, the expression **72** classifies as **integer**, but you could also choose the types **positive** and **posint**, according to Maple.

4.1.2   Checking a Type

How does the **type** function work? To test the classification (*type*) of an expression (*expr*), enter **type(*expr*,*type*)**. Maple will report *true* or *false*, depending on whether or not *expr* classifies under *type*. For example, check if Maple knows that 72 is an integer:

Step 48: The **type** Function

```
> type(72,integer);
```
                                        Check if the expression 72 is an integer.
                    true   *Yes, 72 is an integer.*

You might be wondering how you are supposed to know all the types Maple offers. Look at `?type` to discover the complete list, but don't worry! You do not need to memorize all of the types—just remember how you found the list.

4.2   TYPE CLASSIFICATIONS

Suppose you wish to check the expression $\sqrt{a+b}$. Does this expression classify as addition, function, or a power? You need to guide Maple a bit. Maple classifies expressions in four manners, as in `?type[definition]`. The sections that follow concentrate on the "types of types" that you will frequently encounter early in your studies. See also `?subtype`.

4.2.1   Surface Type

A *surface type* is the uppermost element, or top node, in an expression tree. Since operators may be the top node, Maple includes them as possible surface types. For instance, the surface type of $\sqrt{a+b}$ is a power, which Maple signifies as a ` ^ `. When a type is an operator, you must use backquotes so Maple doesn't think you're trying to use that operator! Other surface types include more general classifications, like **even** and **mathfunc**. For a full list of surface types that Maple recognizes, consult `?type[surface]`.

4.2.2   Nested Type

A *nested type* is a type that Maple looks for within the entire expression. For instance, the expression $a+1$ is not **constant**. However, Maple considers $1+\sqrt{2}$ as **constant** because all the elements in the expression are constant. Maple provides a brief list of nested types along with surface types in `?type[surface]`.

4.2.3 Structured Type

A **structured type** is a type built from individual types in an expression. When expressing a structure type, you replace tokens and expressions with their types. For instance, the structured type of a^2 is **name^integer**. Finding the structured type for $\sqrt{a+b}$ is left as a homework problem. Figuring out a structured type that Maple accepts can be quite challenging sometimes, and you may find the listing of rules **?type[structured]** a bit vague. Fear not—you do not usually need to know structured types unless you need to work on advanced problems.

4.2.4 Finding Types

What if you want Maple to tell you the type of an expression? Maple will tell you the surface type of an expression with **whattype(expr)**. For instance, ask Maple to report the type for **a+b*c**:

Step 49: Reporting Surface Type

```
> restart;
```
Refresh your worksheet.

```
> whattype(a+b*c);
                    +
```
Query Maple about the type for $a + bc$.

Maple reports the surface type +.

Why did Maple report the addition operator (**+**)? In the expression tree, the plus sign is the top node that connects the subexpressions **a** and **b*c**, as shown in Figure 4.1. For more information and related functions, consult **?whattype**, **?op**, and **?hastype**. Now that you know how to find and check types, you will explore the types Maple offers throughout the rest of this chapter.

PRACTICE!

1. Does Maple consider the surface type of the expression $\sin(x)$ as **function**?
2. How do you check the type for the expression $a + b$? Hint: Investigate **?type[arithop]** and **?arithop**.
3. How might the expression types **anything** and **type** help you test expressions?
4. Enter **whattype(1+2)**. Why is the expression type not **+**?

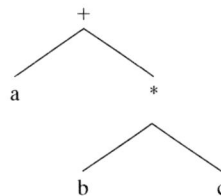

Figure 4.1 Expression Tree for $a + bc$

4.3   REAL NUMBERS

Most forms of engineering and science measure quantities with real numbers. Thankfully, Maple enumerates numerous numerical number types that are reviewed by this section. Experienced users seeking detailed information on numerical computations (with less alliteration) should investigate **?numerics**.

4.3.1   Finding Number Types

When you want to determine the type of a number in Maple, use **NumericClass (expr)**. Maple will attempt to identify the classifications that **expr** might fit. For example, Maple knows that 10 is a positive integer:

Step 50: Identifying Number Types

```
> NumericClass(10);
                    posint
```
Query Maple about the type of the expression 10.
Maple determines that 10 *is a positive integer.*

All numerical types described in this chapter are known by **NumericClass**. For a full listing of numeric types, consult **?numeric_types**.

4.3.2   Integers

Integers are exact, whole numbers written as a series of digits with no decimal point.

- *Natural integers* are strictly positive.
- *Signed integers* have positive or negative signs.

For example, 72 and 1216 are integers, whereas 72.0, 0.01, and $\sqrt{2}$ are not integers. Maple prefers integers because integer operations produce exact results:

Step 51: Integers

```
> 1+10+100;
                    111
```
Add integers together. Do not use decimal points.
Maple retains exact values whenever possible.

Consult **?integer** and **?type[integer]** for more information. You will find integer-related functions in Mathematics...Numbers...Integer Functions....

4.3.3   Fractions

Maple defines **fractions** as numbers of the form $\dfrac{signed\ integer}{natural\ integer}$, where the signed integer is divided by the natural integer. You may also call the "top value" the *numerator*, and the "bottom value" the *denominator*. Because a fraction involves division, enter a fraction using the syntax **numer/denon**. If you give a negative denominator, Maple will automatically move the minus sign to the numerator to maintain fractional form:

Step 52: Fractions

```
> 2/(-3);
                    -2
                    --
                     3
```
Divide two integers to produce a fraction.

Maple converts the expression to fractional form.

Why does Maple leave a fraction undivided? Maple strives to maintain exactness whenever possible. Thus, expressions that cannot be automatically simplified are left untouched. For more information and related functions, consult **?fraction** and **?type[fraction]**.

4.3.4 Rational Numbers

Rational numbers include both integers and fractions. The word *rational* arises from the notion of *ratio* of values. Maple automatically simplifies rational numbers by removing common factors. As with fractions, Maple will maintain exactness and avoid dividing fractions that might introduce irrational results. In the following example, Maple can use automatic simplification without any problem:

Step 53: Rational Numbers

```
> 1/3 + 1/3 + 1/3 - 1;
                    0
```
Add rational numbers together.
Maple simplified the expression.

For more information and related functions, consult **?fraction** and **?type[rational]**.

PRACTICE!

5. Does Maple simplify the expression $\frac{72}{42}$? If so, what is the result?
6. Will entering **1./3.+1./3.+1./3.** produce *integer* output 1? Why or why not?

4.3.5 Floating-Point Numbers

Since the "real world" cannot always provide exact numerical values, numerical analysis relies on **floating-point numbers** (or just *floats*) to measure quantities. Floats are base-10 or *decimal* numbers, like 11., 10.1, and 0.01. Maple considers a number input with a decimal point (**.**) as a float, but you may enter floats in other ways, as shown in Table 4.1.

When using **E**, you may interchange **e** for **E**. The **E** and **e** notation abbreviates the function **Float(x,y)**, which represents a float in scientific notation $x \times 10^y$, given *mantissa x* and *exponent y*. When reporting "big" numbers, Maple will show them in scientific notation without the multiply symbol ($\times$) in the output. For instance, enter 1 million, where $1,000,000 = 1 \times 10^6$:

TABLE 4.1   Entering Floats

Syntax	Example Input	Example Output
int.int	100.1	100.1
.int	.01	.01
int.	10.	10.
*int.int*E*int*	1.2E3	1200.
*int*E*int*	1E2	100.
Float(x,y)	Float(1.2,-3)	.0012

Step 54: Scientific Notation

> **Float(1,6);**   *Enter 1 million. You could also enter 1E6 or 1e6.*

$$0.1\ 10^7$$ *Maple calculates* $1 \times 10^6 = 0.1 \times 10^7$.

Have you wondered why floats aren't included as Maple tokens? Maple treats all floats that you enter as a pair of integers, mantissa **x** and exponent **y**, using scientific notation. For more information on floats and additional background, consult **?float**, **?type[float]**, and **?type[numeric]**.

4.3.6 Floating-Point Arithmetic
Arithmetic operations with floats produce floats:

Step 55: Floats

> **Float(-1,2) - 20. - 3.0 - 0.4 - 5E-2 - 6.0E-3 - 7e-4;**

$$-123.4567$$ *The floats above demonstrate the variety of forms.*

Maple automatically converts integers to floats when types are mixed:

Step 56: Mixing Floats with Integers

> **0.5 + 1/2;**   *Add the float 0.5 to the rational number $\frac{1}{2}$.*

$$1.000000000$$ *Mixing floats and rational numbers produces floats.*

Beware that mixing floating-point numbers with other types tends to force all evaluations to decimal form.

PRACTICE!

7. Express 123.0 in Maple floating-point notation with three different syntaxes.
8. Why does entering **123*10^(-1)** yield a rational number? Change the input to produce a floating-point answer.

4.3.7 Evaluating Floats
Suppose you wish to compute a floating-point result from an integer operation. Remember that Maple keeps calculations exact if possible, so dividing 1 by 2, for example, will not yield a float:

Step 57: Integer Division

> **1/2;**   *Divide 1 by 2.*

$$\frac{1}{2}$$ *Integer division produces a fraction, not a float.*

So, how do you force Maple into producing a float? One option is to enter one of the numbers as float, since mixing a float with integers produces a float:

Step 58: Produce Float Using a Decimal

```
> 1/2.0;
```
$$0.5000000000$$

Divide 1 by the float 2.0.
Mixing floats and integers produces floats.

Or, you may **evalf(*expr*)** ("evaluate float") to force a floating-point computation:

Step 59: Produce Float Using **evalf**

```
> evalf(1/2);
```
$$0.5000000000$$

Evaluate the floating-point result of 1/2.
evalf produces a float.

4.3.8   Numerical Precision

Numerical precision refers to the number of digits to use in floating-point calculations. Realistically, many engineering quantities have uncertain values, so finding the hundredth decimal place for length of a beam, for example, is rather pointless.[1] However, some manufacturing processes do require levels of precision in their measurement, and you can use Maple to help guide those calculations. Maple has two ways of handling numerical output:

- By modifying the display by limiting the number of digits that are shown.
- By limiting the number of digits that are used in the calculations.

By default, Maple uses and shows 10 digits, as shown in previous steps. To change the display and/or Rounding, select File→Preferences... and click the Numerics tab. Display Results modifies only the output display, and I recommend setting this field to 3 or 4 decimal places. Calculation actually sets the number of digits to use, so you should be really sure how many significant figures are actually in your model.

You can provide these features at the Maple prompt. For the display, use the **interface(displayprecision=*value*)**, where **value** is the number of digits to display:

Step 60: Change the Numerical Display

```
> interface(displayprecision=2):
```
Set the display to two digits.

```
> 1.0/2;
```
$$0.50$$

Force a floating-point result.
Maple doesn't count the leading zero as a digit!

To set the maximum number of digits Maple uses, Maple predefines a variable called **Digits**. By default, Maple sets **Digits** to 10, which explains why you have seen upwards of 10 digits in floating-point output. To change the number of digits Maple uses, assign a new value to **Digits**:

Step 61: Changing **Digits**

```
> Digits:=3:
```
Change the number of digits Maple uses to 3.

---

[1] If you pursue science and engineering, you should become comfortable with *significant figures*, which represent the maximum number of digits to use in a calculation. Maple actually helps you with this methodology, as presented in this section.

> `1/2.0;`   *Divide* 1 *by the float* `2.0`.

$$0.500$$ *Maple used only* 3 *digits to compute* `1/2`.

To reset the value of **Digits**, reassign the value of 10 or use the Preferences window again. For further help on these features, see **?preferences**, **?displayprecision** or **?interface**, and **?Digits**.

When you wish to evaluate a single expression, you should choose **evalf(*expr, n*)**, which sets **Digits** to *n* only during the evaluation of *expr*. If *n* is greater than **Digits**, Maple will assume *n* equals **Digits**. Try the following example for practice:

Step 62: Evaluate to n Digits

> `evalf(1/2,3);`   *Evaluate the floating-point result of* `1/2` *to* 3 *digits.*

$$0.500$$ **evalf** *produces a float with only* 3 *digits.*

Digits is actually an ***environment variable***, which is a predefined name that can affect the behavior of Maple during evaluation. You may read more about **Digits** and other environment variables in **?environ** and Appendix D. For further information, consult **?evalf**.

4.3.9   Rounding

What if your input's amount of digits exceeds **Digits**? Maple will approximate the expression by evaluating the float and *rounding* the result up or down. Maple uses an environment variable called **Rounding** to control the direction of rounding. You can check the current value of **Rounding**:

> `Rounding;`   *Check Maple's current rounding scheme.*

$$nearest$$ *Maple sets* nearest *as the default value.*

By default, Maple sets **Rounding** to *nearest*, which forces rounding to the nearest possible value for expressions that contain a number of digits in excess of **Digits**. Maple rounds the digits to the left of the extra digits in the following "directions":

- up, if the extra digits represent a value greater than 5 times 10, raised to a power.
- down, if the extra digits represent a value less than 5 times 10, raised to a power.

For example, suppose you need to round 1,049 to 2 digits using Maple. The 49 rounds down because $49 < 50$, or $4.9 \times 10^1 < 5.0 \times 10^1$.

Step 63: Rounding Mystery Resolved

> `evalf(1049,2);`   *Evaluate* 1049. *to a 2-digit float.*

$$1000.$$ *Maple rounds down because* 49<50.

What happens when the extra digits form a value exactly equal to 5, multiplied by a power of 10? This situation is called a *rounding conflict* or *tie*, which *nearest* resolves "by rounding to the nearest even," according to Maple's Help. In other words, if the digit before the extra 5 is odd, Maple will round the number up. Otherwise, if the digit is even, Maple will round the number down by truncating the 5. For example, assuming you need 3 digits, Maple rounds 1.025 down to preserve the even digit 2:

Step 64: Rounding Mystery Resolved

```
> evalf(1.025,3);
                1.02
```
Evaluate the expression 1.025 *to three digits.*
Maple rounds 1.025 *down because the third digit* 2 *is even.*

What about 1.035? Maple rounds the 3 up o 4 to make an even digit:

```
> evalf(1.035,3);
                1.04
```
Evaluate the expression 1.035 *to three digits.*
Maple rounds 1.035 *up because the third digit* 3 *is odd.*

Maple permits alternative rounding methods listed in **?Rounding**. For example, you could assign **Rounding** to **0**, which forces Maple to round down in all cases. See **?trunc** for related rounding functions.

PRACTICE!

> 9. Evaluate $\frac{1}{3}$ to 4 digits.
> 10. Why does the input **Digits:=2: 1.05+1.05**; produce the output 2.0?

4.3.10 Irrational Numbers

An ***irrational number*** is not crazy! It just cannot be expressed as a ratio of two integers. When expressed as floats, irrational numbers have nonrepeating decimals with no determined bounds. Some examples of irrational numbers include $\sqrt{2}$, π, and the exponential constant e. To maintain exactness, Maple typically will "refuse" to generate a floating-point result and, instead, will leave an irrational quantity in symbolic form. For example, enter **sqrt(2)** to show that Maple knows that $\sqrt{2}$ is irrational:

Step 65: Display Irrational Number

```
> sqrt(2);
                √2
```
Attempt to evaluate $\sqrt{2}$.
Maple cannot further reduce the irrational value $\sqrt{2}$.

To force a floating-point approximation of **sqrt(2)**, you could enter **sqrt(2.0)**:

Step 66: Compute Irrational Number

```
> sqrt(2.0);
           1.414213562
```
Evaluate a numerical approximation of $\sqrt{2}$.
The float 2.0 *forces a floating-point evaluation.*

Or, use **evalf** by entering **evalf(expr)**:

Step 67: Compute Irrational Number

```
> evalf(sqrt(2));
           1.414213562
```
Evaluate a numerical approximation of $\sqrt{2}$.
evalf *forces a floating-point evaluation.*

4.3.11 Pi

A famous—and very useful—irrational number used throughout science and engineering is *Pi*, which has the value 3.1415.... You will often see Pi written as the Greek character π. Maple implements π as the *symbolic constant* **Pi**, which is a predefined name used as a value in expressions. So, Maple will not show a numerical equivalent:

Step 68: Numerical Pi

> `Pi;`   *Maple defines* **Pi** *as the numerical value* π.
    π *Maple automaically displays* **Pi** *as* π.

To show the numerical value, use **evalf**:

Step 69: Evaluate Numerical Pi

> `evalf(Pi);`   *Maple defines* **Pi** *as the numerical value* π.
    3.141592654   *Maple automaically displays* **Pi** *as* π.

Remember that Maple is case sensitive! A common mistake new users make is entering lowercase **pi** instead of titlecase **Pi**, because Maple outputs both as π:

Step 70: **pi** is Not the Numerical Pi!

> `pi;`   *This input is lowercase* **pi**, *not* **Pi**!
    π *The lowercase Greek name* π *is* **pi**.

> `evalf(pi);`   *What happens if you attempt to evaluate* **pi**?
    π *You do not obtain the numerical* **Pi**!

If you use the symbol palette, click the π symbol in the bottom row to get the numerical version.

PRACTICE!

11. Classify the following expressions as rational or irrational: $\sin(0)$, 10.1, $\sqrt{2}$, e.
12. Evaluate π to five digits.
13. Compute $4^{\frac{1}{3}}$ in both exact and floating-point forms.

4.4   COMPLEX NUMBERS

Can you take the square root of a negative number? Technically, no, but some physical quantities are negative due to sign conventions. To resolve this dilemma, imaginary values were conveniently invented to represent square roots of negative numbers that often arise in mathematical analysis. How "imaginary" is an imaginary value? The next section answers that question.

4.4.1   Definitions

To handle negative square roots, the name i is defined as follows:

$$i = \sqrt{-1}. \tag{4.1}$$

Therefore, $i^2 = -1$. So, wherever you see a negative square root, write i instead, as in $\sqrt{-2} = \sqrt{2}i$. In general, a number in the form $\sqrt{-n}$ is written as $\sqrt{n}\,i$ and called a *purely imaginary number*. Adding a purely imaginary number to a real number, as in

$a + bi$, creates a *general complex number*. For instance, to express $4 - \sqrt{-3}$, you would write as $4 - \sqrt{3}i$. The term ***complex number*** refers to both purely imaginary and general complex numbers. Consult **?complex**, **?type[imaginary]**, and **?type[complex]** for more information.

4.4.2 Arithmetic

When you need $\sqrt{-1}$, click i from the symbol palette or enter **I**. Remember, Maple will show i as an uppercase I.

Step 71: Complex Number **I**

> **sqrt(-1);**

$$I$$

Enter i using $\sqrt{-1}$.
Maple represents $\sqrt{-1} = i$ as I.

To create a complex number or general expression in the form $a + bi$, you would enter **a+b*I**. You can use complex expressions in most of Maple, which will attempt to evaluate both real and imaginary components unless you use the **RealDomain** package (see **?RealDomain**). For instance, try one of the arithmetic operations summarized in Table 4.2.

Step 72: Complex Arithmetic

> **(1+2*I) + (2+I);**

$$3 + 3I$$

Add two complex numbers.
$(1+2i)+(2+i) = (1+2)+(2+1)i = 3+3i.$

In earlier releases, Maple implemented **I** as an *alias* (see **?alias**), but now **I** automatically calls **Complex(1)**, which represents i. You may enter an imaginary number **a*I** as **Complex(a)** and a complex number **a+b*I** as **Complex(a,b)**, if you prefer:

Step 73: Complex Numbers using **Complex**

> **Complex(1) + Complex(1,3);**

$$1 + 4i$$

Add two complex numbers.
$(0+1i)+(1+3i) = (1)+(1+3)i = 1+4i.$

For more information, consult **?I** and **?complex**. Adventurous students who want to change Maple's implementation of i should consult **?interface**.

4.4.3 Functions

For functions that relate to complex numbers, check out **?Re**, **?Im**, **?evalc**, and Mathematics...Numbers...Complex Numbers.... These functions are reviewed in later sections.

TABLE 4.2   Complex Arithmetic

Operation	Rule
Addition	$(a + bi) + (c + di) = (a + c) + (b + d)i$
Subtraction	$(a + bi) - (c + di) = (a - c) + (b - d)i$
Multiplication	$(a + bi)(c + di) = (ac - bd) + (bc + ad)i$
Division	$\dfrac{a + bi}{c + di} = \dfrac{ac + bd}{c^2 + d^2} + \dfrac{bc - ad}{c^2 + d^2}i$

PRACTICE!

14. Do the inputs **sqrt(-1)**, **(-1)^(1/2)**, and **I** produce different outputs?
15. Evaluate $(1 + 2i)(-1 - i)$.

4.5 EXTENDED NUMERICS

This section describes some handy commands that relate to numerical analysis in Maple. Consult **Mathematics...Numerical Computations...Maple Numerics...** for a broad listing of numerical features that Maple offers.

4.5.1 Infinity

Many calculations require that you use a really, *really* big positive number called *infinity*, denoted as ∞. Maple's extended numerics consist of the ability to represent ∞ with the symbolic constant **infinity**:

Step 74: infinity

> **infinity;**   *Evaluate the symbolic constant* **infinity**.
                                            You may also use the symbol palette to enter ∞.
$$\infty$$

Of course, adding a number to ∞ cannot make ∞ any bigger, since ∞ is already the biggest thing imaginable. Maple agrees!

Step 75: Operations with infinity

> **infinity + 32;**   *Evaluate* $\infty + 32$.
                                          Maple can perform operations with ∞.
$$\infty$$

For more information, consult **?infinity**, **?type[infinity]**, and **?type[numeric]**.

4.5.2 Example

Why should you be concerned with infinity? One handy application is the generation of general plots without having to specify numerical ranges. For instance, if you wish to investigate the behavior of the square function $\sqrt{x}$ for all positive numbers, you could plot the function for the range $0 \leq x \leq \infty$. In Maple, you would do the following:

Step 76: Plotting Infinity

> **plot(sqrt(x),x=0..infinity);**   *Plot* $\sqrt{x}$ *for* $0 \leq x \leq \infty$.

The "infinite" plot shows general trends.

4.5.3   Undefined

As an amusing thought exercise, consider what subtracting infinity from infinity might mean. Should the result be zero, and if so, what guarantee do you have that a really big number is the same size as another? Maple sidesteps this issue by using the predefined name **undefined** for operations that cannot produce a number:

Step 77: undefined

> **infinity-infinity;**   *Attempt to evaluate* $\infty - \infty$.

undefined   *Maple reports* **undefined** *as a "value."*

Maple classifies numbers that contain **infinity** and/or **undefined** as *extended numerics*. The **RealDomain** package will produce **undefined** instead of complex results. Chances are you still won't confront **undefined** too often, but it does arise if you perform unrecognized operations with ∞. You may also actually use **undefined** as a kind of value that you operate on, as discussed in **?undefined** and **?type[undefined]**.

4.5.4   Complex Extended Numerics

Maple describes handling infinite complex numbers in **?complex**.

PRACTICE!

16. How does Maple treat division by zero?
17. Plot e^{2x} for $-\infty \le x \le \infty$.

4.6   BOOLEAN EXPRESSIONS

Maple provides more than just numerical tools—you also can create expressions that incorporate logic and comparison relationships. These expressions are called **Boolean** because they deal with the truth of an expression. Boolean expressions use two other types, *logical* and *relational*, which are described in this section.

4.6.1   Logic

You may model the truth of an assertion with logic. For instance, the assertion *this book teaches Maple* is (I hope) true, whereas the statement *this book teaches spelunking* is false. Maple implements the logical values *true* (T) and *false* (F) as symbolic constants, **true** and **false**, respectively. Maple also uses a third value, the symbolic constant *FAIL*, which is returned when Maple cannot determine the truth of an expression. For instance, the statement *this book blurgles foogly-woo* is a good candidate for *FAIL*.

4.6.2   Logical Expressions

You can create *logical* expressions by operating on symbolic expressions and other Boolean expressions with the operators **and**, **or**, **not**, **xor**, and **implies**. See Table 4.3 for common Standard Math notation for some of these operators.

TABLE 4.3   Logical Operators

Operator	Standard Math	Maple Notation
and	$\wedge$	**and**
or	$\vee$	**or**
not	$\neg$	**not**

For instance, suppose you want to store the expression $a \wedge b$ in the variable x:

Step 78: Logical Expression

```
> restart;
```
Refresh your worksheet.

```
> x := a and b;
```
$$x := a \text{ and } b$$
Assign the expression $a \wedge b$ to x.
Remember that and is an operator.

What can you do with a logical expression? Logical operators act upon the logical values *true* and *false*. Suppose you assign *true* to a and to b. Logic dictates that $true \wedge true = true$, and thus, $a \wedge b = true$. For example, you can test if Maple considers the number 10.1 both floating-point and numeric by using the **type** function, which evaluates to *true* or *false* when testing an expression's type:

Step 79: Using a Logical Expression

```
> a := type(10.1,float):
> b := type(10.1,numeric):
```
Assign to a the result of checking if 10.1 is a float.
Assign to a the result of checking if 10.1 is numeric.

```
> x;
```
$$true$$
Evaluate x, which stores $a \wedge b$.
*Maple reports that $a \wedge b$ evaluated to **true**.*

In the preceding step, Maple evaluated a as *true* and then b as *true*. Then, after you entered x, Maple evaluated x as a **and** b, which evaluated to $true \wedge true$, which simplified to *true*. Table 4.4 summarizes the rules for combining *true* (T) and *false* (F). Refer to **?boolean** for the full tables that include *FAIL* for all of Maple's logical operators.

4.6.3 Relations

Logical expressions do not compare numerical quantities. To do so, use a ***relation***, which is an expression composed of two subexpressions and a relational operator. Typically, relations involve comparisons, like the expression $3 < 4$. In general, a relation compares two expressions ***expr1*** and ***expr2*** with an operator ***rel*** using the syntax ***expr1 rel expr2***. Table 4.5 summarizes Maple's relational operators.

Maple automatically reverses "greater than" inequalities into "less than" form.[2] For example, try storing the relation $1 > 0$ in the name *REL*:

Step 80: Store Relation

```
> REL := 1>0;
```
$$REL := 0 < 1$$
Store the expression $1 > 0$ inside REL.
Maple automatically converts ">" into "<" form.

TABLE 4.4 Logical Operations

$\wedge$	T	F		$\vee$	T	F		$\neg$	
T	T	F		**T**	T	T		**T**	F
F	F	F		**F**	T	F		**F**	T

---

[2] Why bother? According to **?equation**, "inequalities involving the operators **>** and **>=** are converted to the latter two cases [**<** and **<=**] for purposes of representation." I am still unsure why this is important, other than saving the user from memorizing two more types.

TABLE 4.5   Relational Operators

Operator	Standard Math	Maple Notation	Example
Equal	$=$	**=**	**1 = 1**
Not Equal	$\neq$	**<>**	**1 <> 2**
Greater Than	$>$	**>**	**2 > 1**
Greater Than or Equal	$\geq$	**>=**	**2 >= 1**
Less Than	$<$	**<**	**1 < 2**
Less Than or Equal	$\leq$	**<=**	**1 <= 2**

Maple stored the equivalent relation $0 < 1$, instead. Note that *REL* now refers to the entire relation $0 < 1$:

Step 81: Check Relation

> **`type(REL,relation);`** *Check if REL stores an expression of type **relation**.*

$$true$$ *Yes, REL is a relation.*

You will see how to evaluate a relation in Section 4.6.5. For more information about relations and their operators consult **`?inequality`** and **`?operators[binary]`**.

4.6.4   Equations

An *equation* is a relation that uses the equals operator (**=**). So, an equation has the syntax **`expr1 = expr2`**. Remember that the equals sign (**=**) does not assign expressions! Instead, use equations for comparison and solving, not for assignment:

Step 82: Assign Equations

> **`restart;`** *Refresh your worksheet.*

> **`EQN := y=m*x+b;`** *Assign an equation to name EQN.*

$$y = mx + b$$ *EQN, and not y, stores the equation.*

> **`EQN;`** *Show the value of EQN.*

$$y = mx + b$$ *Yes, EQN really has the value y = mx + b.*

> **`y;`** *Show the value of y.*

$$y$$ *The equals sign does not assign!*

Maple uses equations very often. For instance, to solve for x in the equation $y = mx + b$, you could use **`solve(eqn,var)`** to solve for variable **`var`** in equation **`eqn`**:

Step 83: Solving Equations

> **`solve(EQN,x);`** *Solve y = mx + b for x.*

$$\frac{y - b}{m}$$ $x = \frac{y-b}{m}.$

*x is not assigned unless you enter **`x:=solve(EQN,x)`**.*

Equations are discussed further in **`?equation`** and **`?type[equation]`**. See **`?solve`** for more information about solving equations.

4.6.5 Boolean

Any expression composed of logical and/or relational subexpressions is called *Boolean*. Boolean expressions evaluate to *true, false,* and *FAIL,* using rules of logic. Maple performs a Boolean evaluation in the following situations:

- An expression contains a logical operator: **and, or, not, xor, implies**.
- You perform a test inside an **if** or **while** clause: See Appendix D.
- You enter **evalb(*expr*)** to evaluate the Boolean value of *expr*.

Try the following steps for some examples of Boolean expressions:

Step 84: Boolean Evaluation

```
> 1 > 2 or 1 < 2;
```
                                    Ask Maple to evaluate if 1 *is greater than or less than* 2.
              true   *Maple evaluates your Boolean expression as* **true**.

```
> evalb( 1 > 2 );
```
                                    Evaluate the Boolean value of the expression 1 > 2.
              false   *Maple evaluates your Boolean expression as* **false**.

When using **evalb(*expr*)**, beware of the following:

- Maple will not simplify *expr* except for automatic simplifications.
- **evalb** does not assign expressions and will produce only *true, false,* or *FAIL*.

For information, consult **?boolean, ?evalb, ?type[boolean]**, and **?FAIL**.

PRACTICE!

18. Evaluate $\neg\,((\mathit{true} \lor \mathit{false}) \land \mathit{true})$.
19. Solve for x in the equation $ax^2 + bx + c = 0$, using the **solve** function.
20. Evaluate the truth of the relation $10 + 2 \geq 12$.
21. Does Maple automatically "know" that $x^2 - 1 = (x + 1)(x - 1)$? Why does entering **evalb((x^2-1)/(x-1) = (x+1))** evaluate to *false*?

4.7 MULTIPLE EXPRESSIONS

Many models require you to analyze and design parameters for different values. For situations in which you need to collect expressions together, Maple provides many types that can represent multiple expressions. This section reviews common types. For more information, refer to **?exprseq**.

4.7.1 Ranges

A ***range*** is a continuous interval of values between, and including, the endpoints. You may express a range as r and the endpoints as $[a, b]$. You may also think of r as all values of a variable x such that $a \leq x \leq b$ is true. Maple represents the range r as the expression ***a..b***, using the range operator **..** (double-dot). You will commonly find continuous ranges in plotting:

Step 85: Continuous Ranges

```
> plot(x^2,x=0..100):
```
                                    Plot x^2 *for the range* $0 \leq x \leq 100$.
                                    Plot not shown to conserve space.

4.7.2   Sequences

A "discrete range" has a more appropriate term called **sequence**, which is a collection of items placed in a specific order. In terms of Maple, I'm technically "breaking the rules" a bit by introducing *expression sequences* in this chapter because they are *not* types! Actually, a sequence is a collection of expressions in a specific order. To create a sequence of expressions, separate each expression with a comma (**,**), as in **expr1, expr2, ...** Each expression is an *element* of the sequence. For example, try storing the sequence of expressions 1, 2, $\sqrt{2}$ in a variable *Seq*:

Step 86: Discrete Ranges

> **Seq:= 1, 2, sqrt(2);**   *Store an expression sequence in* Seq.
$$Seq := 1, 2, \sqrt{2}$$ *Seq holds all the values in the specified order.*

You may also generate a range of discontinuous, or *discrete*, values using the **$** operator with the syntax **$a..b**:

Step 87: Discrete Ranges

> **$1..3;**   *Generate integers between, and including* 1 *and* 3.
$$1, 2, 3$$ *Maple produces a non-continuous range.*

For more information about further syntax and applications of the range operator, consult **?range**, **?plot[range]**, **?type[range]**, and **?$**.

  You might get confused if you enter **whattype(Seq)**, because Maple will report *exprseq*, which seems to imply that *exprseq* is a type. But, if you enter **type(Seq, exprseq)** Maple will generate an error message! Why? Elements of a sequence may have different types, so Maple wants to "avoid" possible confusion. For further explanation, consult **?type** and the overview of sequences in **?exprseq**. You will find information about related material in **?seq**, **?$**, **?map**, and **?op**.

4.7.3   Lists

A **list** is an expression that contains a sequence of expressions. To create a list, enclose a sequence with square brackets (**[]**). For example, suppose you wish to collect a list of numbers and names in a list called *List*:

Step 88: Lists

> **List := [x1,23,-3.0,x2,x1];**   *Assign a list by surrounding a sequence with (* **[]** *).*
   $$List := [x1, 23, -3.0, x2, x1]$$ *You may mix expression types in a list.*

As in the preceding example, you may repeat list elements. For more information and related functions, consult **?list**, **?type[list]**, **?op**, **?map**, and **?ListTools**.

4.7.4   Sets

A **set** is an expression that collects *unordered unique* items. To create a set, surround a sequence with curly braces (**{}**). In the following example of assigning a set, Maple will remove duplicated elements to maintain each element's uniqueness:

Step 89: Sets

```
> Set := {e1,e2,e2,e3};                          Assign to the name Set a set.
                    {e1, e2, e3}        Maple removes repeated elements in creating a set.
```

For information about sets and related operations, consult **?set**, **?type[set]**, and **?union**.

4.7.5   Selection

Besides expressing lists, square brackets (**[]**) assist with indexing and extracting elements from sequences, lists, and sets. Maple counts each expression inside a sequence, list, and set from left to right starting from 1. Given a sequence input **Seq**, entering the expression **Seq[i..j]** extracts the ith to jth expressions from **Seq**. To extract just one element at the ith position, enter **Seq[i]**:

Step 90: Selection

```
> restart:                                                Refresh the worksheet.

> Seq:=a,b,c,d,e:                              Assign to Seq the sequence a,b,c,d,e.

> Seq[1..3]; Seq[5];     Extract the first through third elements. Then, extract the fifth element.
                  a, b, c         Maple extracted the 1st, 2nd, and 3rd elements from Seq.
                      e                    Maple extracted the 5th element of Seq.
```

For an extensive overview, consult **?selection**.

PRACTICE!

22. What does the input statement **'i^2' $ 'i' =0..3** indicate and produce? Hint: Consult **?$**.
23. Produce the sequence of integers 0, 1, 8, 27. Assign the result to S.
24. Show the first and second elements of S.
25. Create a list using S. Assign the list to SL.
26. Create a set from S. Assign the set to SS.
27. Does Maple consider SL and SS identical?

4.8   NAMES

Maple names act as variables that can store expressions. Since a name can represent an expression, a name must also be an expression. This section reviews further aspects of Maple names.

4.8.1   Symbols and Names

According to **?name**, a *symbol* is an expression composed of a letter followed by zero or more letters, digits, and underscores. A *name* may be either type **symbol** or **indexed**, which means a name can have an index. Maple refers to names with indices as *indexed names*.[3]

---

[3]Actually, I think that this term is redundant and that *indexed symbol* would be more appropriate.

4.8.2   Indexing

Indexing is the process of labeling a particular expression with an *index* or *subscript*. In Maple, an index may be any sequence of expressions. When using Maple Notation, use square brackets to indicate an indexed expression, as in **expr[index]**. When **expr** is unassigned, Maple will report **expr** with **index** as a subscript:

Step 91: Indexing

> **x[1],x[2],x[3];**   *Provide indices for x.*

x_1, x_2, x_3 *Because x has no assignment, no elements are extracted.*

Why bother? Indexing allows you to provide very descriptive names in your output without clunky variable names, like **x1** and **x2**.

If the name stores multiple expressions, as in a range, sequence, list or set, Maple will extract the expression at the position indicated by the index, as discussed in Section 4.7. For instance, given a list $L = [a, b, c, d, e]$, the notation L_1 refers to element a, L_2 refers to element b, and so forth. To refer to any element inside L, you would say L_i. Consult **?indexed**, **?type[indexed]**, and **?op** for more information.

4.8.3   Initially Known Names

Maple predefines many names that are listed in **?ininames**. You will encounter many, such as *true*, *false*, **Digits**, **I**, and **Pi**. The variety is listed below:

- **Symbolic constants**: predefined names that store commonly used values, like *true* and *false*. You may discover the entire collection by entering **constants** at the prompt. See also **?constant** and **?type[constant]**.
- **Environment variables**: names that affect Maple's behavior, like **Digits**. See **?envvar** and Appendix D.

Function names are also protected/predefined names. See **?inifcns** and **?index[function]**.

4.8.4   Greek Names

Maple automatically converts names of Greek characters into a symbolic font. You will find most names also using the symbol palette by selecting <u>V</u>iew→<u>P</u>alettes→<u>S</u>ymbol Palette. Try the following example:

Step 92: Greek Names

> **alpha,beta,gamma,delta;**   *Enter some Greek letters using Maple notation.*

$\alpha, \beta, \gamma, \delta$ *Maple automatically converts each name.*

Beware that Maple uses many uppercase Greek letters for functions, so their names might be protected. For a full list of all Greek symbols, see Appendix A, **?greek**, and **?symbolfont**.

PRACTICE!

28.   What Maple input generates the output $a_b + c_d$?
29.   Generate the full list of symbolic constants (*false*, γ, ∞, *true*, *Catalan*, *FAIL*, π) without entering the individual names.
30.   Assign the value 1 to the name β.
31.   Can you assign the value 1 to the name **Beta**?

4.9 MISCELLANEOUS

Maple expressions include more than numbers and constants. This section only scratches the surface (so to speak) of more Maple types. You will discover these types throughout later chapters and on your own. See also Appendix D.

4.9.1 Arithmetic

As shown in Sections 3.3.3 and 4.2.1, some expressions have arithmetic operators as the "top" node in their trees. Maple classifies such expressions as arithmetic types, using the tokens `'+'`, `'*'`, and `'^'`. For example, the expression **a+b** has the type `'+'`. For more information, consult **?type[arithop]**.

4.9.2 Strings

Strings provide text labels for commenting. Some functions, like **plot**, use strings for labels and titles. As briefly introduced in Table 3.1, to create a string, surround characters with double quotes (**"**). Strings act as "clumps" of characters that you can manipulate and move around. For instance, you may concatenate ("stick together" or append) two strings with **cat(*string1,string2*)**:

Step 93: Concatenating Strings

```
> Str1:="ab":                    Assign to Str1 the string "ab".
> Str2:="cd":                    Assign to Str2 the string "cd".
> String:=cat(Str1,Str2);    Append "cd" to "ab" and assign result to String.
         String := "abcd"            The concatenated string.
```

For more information and examples, consult **?string**, **?type[string]**, the worksheet **?examples[string]**, and **?StringTools**.

4.9.3 Functions

Functions are also expressions and, consequently, one of Maple's most common types. Don't worry, I haven't forgotten them! Functions have the form **expr(*exprseq*)** and are discussed throughout this text.

4.9.4 And More...

Consult Appendix D and **?type** for many other useful types. Many of these types will assist you in developing programs in Maple.

4.10 APPLICATION: OPTIMIZATION

The brief application that follows demonstrates how the variety of Maple expressions can help you along the way as you solve problems.

4.10.1 Problem

Suppose you are designing a rectangular room with only enough supplies to handle 100 m^3. Given a fixed ceiling height of 1 m, what dimensions of the room provide the maximum perimeter?

4.10.2 Background

A rectangular box has *volume = length × width × height*, *area = length × width*, and *perimeter = 2 × (length + width)*. So, the required area is 100 m^2. Now, you need to solve the equations.

4.10.3 Methodology

To solve $100 = length \times width$ and $perimeter = 2(length + width)$ simultaneously, you need to set up the equations with their variables and data. To set up each equation, use the equals operator (**=**), which creates the equation type:

Step 94: Optimization—Restate, Model

> `restart:`   *Refresh your worksheet.*

> `Equ1 := Area = Length * Width:` $A = LW$
> `Equ2 := Perim = 2*(Width+Length):` $P = 2LW$

> `Area := 100:`   *Area has an initial value.*

You can directly solve for **Width** and **Length** with the **solve** function, which has one form syntax of **solve(equs,vars)**, where both **equs** and **vars** are sets. Actually, because the equation for **Area** involves multiplication of variables, you need to apply **allvalues** to force Maple to give explicit solutions:

Step 95: Optimization—Generate Equations

> `Sols := allvalues(solve(Equ1,Equ2,Length,Width));`   *Find L and W.*

$$Sols := \left\{ Length = \frac{Perim}{4} + \frac{\sqrt{Perim^2 - 1600}}{4} \right.$$

$$\left. Width = \frac{Perim}{4} - \frac{\sqrt{Perim^2 - 1600}}{4} \right\},$$

$$\left\{ Length = \frac{Perim}{4} - \frac{\sqrt{Perim^2 - 1600}}{4}, Width = \frac{Perim}{4} + \frac{\sqrt{Perim^2 - 1600}}{4} \right\}$$

As you would expect for a rectangle, the mathematical decision of which side is length and width is irrelevant. Consequently, the solution set confirms this notion, since both solutions are essentially identical. So, you only need to select one, say the first, using set extraction:

Step 96: Optimization—Separate Solutions

> `MySols := Sols[1]:`   *Extract out the first set of solutions from* `Sols`.

Now, you can use a trick in Maple that will actually assign the values of **Length** and **Width** from the set of equations in **MySols**:

Step 97: Optimization—Separate Variables

> `assign, (MySols):`   *Assign* `Length` *and* `Width` *variables.*

> `Length; Width;`   *Show the values of* `Length` *and* `Width`.

$$\frac{Perim}{4} + \frac{\sqrt{Perim^2 - 1600}}{4}$$ *Length*

$$\frac{Perim}{4} - \frac{\sqrt{Perim^2 - 1600}}{4}$$ *Width*

If you find that your answers appear "backward" from mine, remember that sets do not imply order. The swapping means that Maple chose the other solution, which does not matter since both solutions are effectively the same.

4.10.4 Solution

There are several ways to find the range of values of perimeter. I will focus on an approach that uses algebra to continue to demonstrate how understanding types can help you. Since physical lengths cannot be negative, you can solve for values of perimeter that maintain legal values for **Length** and **Width**. Again, I use **solve**, which can solve inequality relations, as well:

Step 98: Optimization—Range of Solution

> **solve(Length >= 0, Perim);**

Find the range of Perim for

$$\frac{Perim}{4} + \frac{\sqrt{Perim^2 - 1600}}{4} \geq 0.$$

$$RealRange(40, \ \infty) \qquad\qquad 40 \leq perimeter \leq \infty$$

> **solve(Width >= 0, Perim);**

Find the range of Perim for

$$\frac{Perim}{4} - \frac{\sqrt{Perim^2 - 1600}}{4} \geq 0.$$

$$RealRange(40, \ \infty) \qquad\qquad 40 \leq perimeter \leq \infty$$

For both cases, Maple produces the same solution RealRange(40, ∞) that is a special kind of continuous-range type, which means $40 \leq perimeter \leq \infty$. Thus, the minimum value of perimeter is 40 m². Using a bit more Maple, you can substitute the value of perimeter with **eval(expr, eqn)**, where **eqn** has the form **oldvar=newvar** and **expr** contains **oldvar**:

Step 99: Optimization—Report Solution

> **eval(Length,Perim=40);**   *Find the length for perimeter of* 40.

$$10 \qquad\qquad length = 10m$$

> **eval(Width,Perim=40);**   *Find the length for perimeter of* 40.

$$10 \qquad\qquad width = 10m$$

Did you expect 10 m for both dimensions? Can you think of a reason why these solutions provide a minimum perimeter? I have used some functions you might not have seen yet. Depending on how readings from this text have been assigned, you will see them soon!

PROFESSIONAL SUCCESS: UNITS

Many programs allow numerical input without unit labels. However, engineering and science quantities measure *units*, which are properties associated with a physical system. Always write your units, especially to help those who check your work—your instructors/managers will thank you. An exciting new feature of Maple is its support of units, as discussed in **?units**. You might find the amount of material there daunting, so read Appendix C where I cover it in more detail. When using the Units package, label your input, and Maple will take care of the conversions. For example, suppose you need to compute the area of a rectangle with the dimensions 10′ × 10′5″ (10 feet by 10 feet, 5 inches). First, restart your worksheet and activate Maple's foot-pound-second (FPS) unit system:

> **Units[UseSystem](FPS):**   *Instruct Maple to use English units.*
> **with(Units[Natural]):**   *Maple will allow you to use unit-labels in your input.*

Ignore Maple's greatly-detailed warning about the functions it has modified.

To use units with your input, multiply each number by the unit label:

```
> side1 := 10.0*ft: side2 := 10*ft+5*inch:
```
Assign the sides using unit labels.

If you investigate **side2**'s value, you will see that Maple automatically converted it to feet. When you find the area, Maple will still keep track of units for you:

```
> area := evalf(side1*side2,3);
```
$$area := 104.[ft^2]$$

Compute the area to 3 digits.
Maple will keep track of units for you!

KEY TERMS

Boolean	complex number	environment variable
equation	extended numerics	floating-point number
fraction	indexing	infinity
irrational number	list	logical
nested type	Pi	range
rational number	relation	sequence
set	structured type	surface type
symbolic constant		

SUMMARY

- Maple classifies expressions according to type.
- For a full list of Maple types, see **?type**.
- A surface type is the top node in an expression tree.
- Maple attempts to preserve exactness in all numerical operations.
- Mixing floating-point values tends to breed further floats.
- For exactness, Maple does not produce floats for irrational numbers.
- You may set rounding with **Rounding** and **Digits**.
- Maple represents imaginary components of complex numbers with **I**.
- You may represent infinite quantities with the symbolic constant **infinity**.
- To represent true and false expressions, use Boolean values and relations.
- Sequences, lists, and sets store multiple expressions.
- Sequences are not formal expressions, whereas lists and sets are.
- Use the **[]** operator to index a name or extract an element from a sequence, list, and set.

Problems

1. What is a *type* in Maple?
2. Explain the difference between surface and structured types.
3. Identify the surface and structured type of the expression $a + bc$.
4. Explain the difference between real and complex numbers.
5. Explain the difference between rational and irrational numbers.
6. Explain the differences and similarities between sequences, lists, and sets.

7. Evaluate the expressions that follow using Maple. You must demonstrate both exact and floating-point (decimal) results for all output. Hints: Be careful with operator precedence and associativity. To find decimal results, either enter floats in the input or use **evalf(expr)**.

   a. $1 + \dfrac{1}{2} + \dfrac{1}{3}$
   b. $1 + 0.51$
   c. $1 + \sqrt{2}$
   d. $\dfrac{1}{3} + \sqrt{0.1}$
   e. 2π
   f. i^2
   g. $\dfrac{2i + 3}{4i}$
   h. $\sqrt{-10}$
   i. $e^2 + 3$
   j. $e^{\sin(\frac{\pi}{3})}$: Hint, use **exp**, not **e**!

8. Is the number zero (0) an integer? Prove your answer using Maple.

9. Show the Maple input that produces the output $\Re(\aleph) + \Im(\aleph)$. Hints: $\aleph$ is not Greek—it's a Hebrew letter that Maple supports without documentation. $\Re$ and $\Im$ are complex functions for extraction of the real and imaginary components of a number, respectively.

10. Maple's **Digits** variable can often produce surprising results. Consider the following session:

```
> restart:
> 2.0*1.05;
                    2.100
> Digits:=2:
> 2.0*1.05;
                    2.0
```

   Demonstrate the same Maple session. Why do you think that Maple produced 2.0 after you set **Digits** to **2**?

11. Determine the structured type for the expression $\sqrt{a + b}$. Fill in the portions indicated by **???** such that the input produces a result of *true*:

```
> type( ???, `^`(`&+`(???,???),???) );
```

12. Derive the formula for complex division shown in Table 4.2 using Maple. Start with the input for $\dfrac{a + bi}{c + di}$ and use the *complex conjugate* of $a + bi$, which is $a - bi$.

13. Why does entering **evalb(exp(1)=e)** produce *false*?

14. The volume of a sphere is $V = \frac{4}{3}\pi r^3$ for radius r. Do the following tasks:

   a. Given $r = 3$ cm, evaluate V. Be sure to specify units in your assignments!
   b. Find the floating-point value of your answer.

c. Convert your result to inches. 1 in = 2.54 cm. Hint: Maple will not permit **in** as name. Try **?in** to see. Use **inch** instead.

15. Given the data in Table 4.6, answer the following questions:
   a. Assign to X the sequence of x values.
   b. Assign to Y the sequence of y values.
   c. Generate a list of data pairs. Each pair of data must have a list structure in the form $[x_i, y_i]$ as well. Assign to the variable DP your result.
   d. Extract the y value from the DP's third value, DP_3. Hint: Use the selection operator **[]** twice.

TABLE 4.6   Data Points

Index	1	2	3	4	5
x	2	−4	3	1	0
y	−1	5	2	6	1

16. One of DeMorgan's Laws states that $(A \cup B)^c = A^c \cap B^c$. Demonstrate this relationship with Maple. Use two sets $A = \{a, b, c\}$ and $B = \{a, d\}$ along with a *Universe of Discourse* $U = \{a, b, c, d, e\}$. Hint: A complement of a set S, called S^c, is the difference between U and S. See also **?minus**.

17. Electrical engineers often distinguish complex numbers with the letter j to prevent confusion with the symbol for current, i. Change Maple's default imaginary number from **I** to **j**. Next, demonstrate your new name by evaluating $\sqrt{-1}$. Hint: See **?I**.

18. Assume part of a circuit you are analyzing contains a *resistor R* (Ω or Ohm) and *capacitor C* (F or farads) connected together in a series, as shown in Figure P4.18.

   (Capacitors store electrical charge.) *Impedance* measures the effect elements, such as capacitors and resistors have on electrical current. You can calculate this configuration's impedance Z (Ohm) with the formula

$$Z = R - \frac{j}{\omega C}$$

   where ω (rad/s) represents the angular frequency of the voltage source. The symbol j represents an imaginary number. Given $\omega = 400$ rad/s, $R = 5\Omega$,, and $C = 1 \times 10^{-9}$ F:

   a. Evaluate Z using Maple's default imaginary number, variable I.
   b. Evaluate Z using j, as described in Problem 17.

Figure P4.18   Resistor and Capacitor in Series

5

Functions

5.1 MAPLE FUNCTIONS

One of Maple's most common expressions is the *function*. You have already used some of Maple's wealth of predefined functions to assist your work. This section reviews further aspects of Maple functions. Inspect **?inifcn** and **?index[function]** for a multitude of Maple library functions. Also, consult Appendix B for methods on finding, loading, and viewing functions.

SECTIONS

5.1 Maple Functions
5.2 Polynomials
5.3 Trigonometry
5.4 Powers and Roots
5.5 Miscellaneous
5.6 Functional Notation
5.7 Application

OBJECTIVES

After reading this chapter, you should be able to

- Identify the syntax of Maple functions
- Distinguish between function input and output
- Find functions from the Maple libraries
- Simulate functions with assignments
- Use functions for polynomials, trigonometry, logarithms, and more
- Develop customized Maple functions using functional notation

Introductory courses in engineering and science provide tools for later study. However, mathematical tools seem to fixate on apparently unimportant theories and concepts. Many students often complain, "Why are we learning *this*? How on earth will *this* ever help us?" Knowing how underlying theories reflect physical principles helps to motivate study. Consider, for instance, why we really need *functions*. Functions provide realistic physical models, as demonstrated in the following example.

Consider the experiment shown in Figure 5.1. A weight w hangs from an unstretched spring and is slowly released. The final displacement u is then measured and plotted. Different weights yield different displacements, as shown in Figure 5.2. Numerous experiments eventually develop a pattern, which is a line drawn through the individual points in Figure 5.3.

Because w is chosen for arbitrary values, w is called an *independent variable* and is plotted on the horizontal axis. Independent variables are chosen and not measured during the experiment. The displacement u depends on the choice of w. Experimentally measured variables are called *dependent variables* and are plotted on the vertical axis.

Each weight causes a unique displacement. Thus, u is a *function* of w. In mathematical terms, $u = f(w)$. From Figure 5.3, the relationship appears to be linear. So, you may express the slope as $\frac{1}{k}$, where k is the stiffness of the spring. Using k, you may express u as $u = \frac{w}{k}$ or just $w = ku$.

Why choose a function? Consider the fake model shown in Figure 5.4. This model predicts that one weight can yield *different* displacements! As a test, you can check if a vertical line will cross the line only once. This fake model fails that test, and thus, is not a function.

The spring example shows that rules for defining functions are crucial for building accurate and realistic models. Now, consider other mathematical theories. Very often mathematical definitions and theories grow from desires to solve certain problems. Though the reasons might seem obscure, have faith that many of your early studies truly have importance!

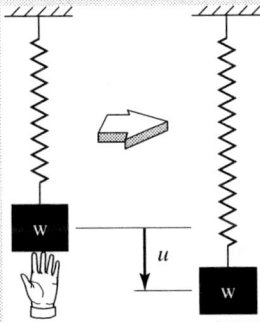

Figure 5.1   Experiments

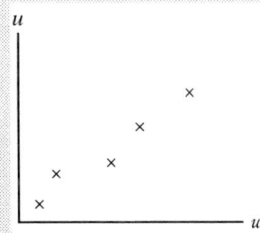

Figure 5.2   Measured Results

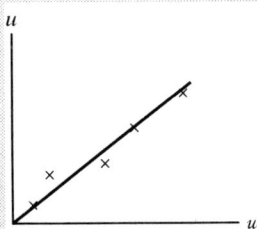

Figure 5.3   Ideal Results

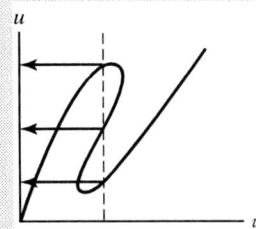

Figure 5.4   Fake Results

5.1.1 Terminology

A *Maple function* has the general syntax **expr(exprseq)**. Although you can make weird-looking functions (see some examples in **?function**), you typically should use the following rules:

- **expr** will be a name that labels the function.
- **exprseq** will be an expression sequence **expr1, expr2**....

Each **expr** inside **exprseq** is called an *argument* or *parameter*. Consult **?function** and **?type[function]** for further discussion.

5.1.2 Input/Output

Arguments serve as function *inputs*, which are quantities upon which the function acts. Some functions, such as **plot(x^2,x=1..2)**, require multiple arguments, whereas functions like **anames()** can accept zero arguments. Generally, functions produce expressions as a result of evaluation. For instance, entering **sqrt(4)** produces the result of 2 from the input 4. You may assign the result of a function's evaluation to a variable:

Step 100: Output from a Function

> **Result := sqrt(4);**   *Evaluate* $\sqrt{4}$ *and store the result in* Result.

$$Result := 2$$ *Maple reports the evaluation and assignment.*

Sometimes programmers refer to evaluating a function as *calling* or *invoking* a function. This terminology derives from thinking of the inputs as messages that you pass when making the "call." Then, the function returns your "call" with an output value. In Step 100, the output was assigned to a name for later use.

5.1.3 Nested Functions

A function is one of Maple's expression types. Since functions have the syntax **expr(exprseq)**, you can write functions as arguments of other functions, which produces *nested functions*. Why bother with nesting? Well, you may enter expressions the long way, as in Step 101:

Step 101: Tedious Method for Many Functions

> **A:=Pi/6: B:=sin(A): sqrt(B);** *Solve for the square root of the sine of Pi divided by 6.*

$$\frac{1}{2}\sqrt{2}$$

Maple evaluated **sqrt(sin(Pi/6))**.

To save time typing, you can nest the functions. For example, **fun1(fun2(expr))** supplies the fully evaluated result of **fun2(expr)** as the input for **fun1**. Then, Maple fully evaluates **fun1** to produce the result for the whole expression. For the preceding example, write **sin(Pi/6)** as an argument of **sqrt**. You should use extra white space to clarify the nesting in case you have a typo:

Step 102: Nested Functions

> **sqrt(sin(Pi/6));**   *Enter* **sin(Pi/6)** *as an argument of* **sqrt**.

$$\frac{1}{2}\sqrt{2}$$ *You get the same answer with less work!*

Maple permits nesting of functions to many levels, as in ***fun(fun(fun(...)))***. When doing so, ensure that your parentheses always match; that is, each open parenthesis (**(**) should have a companion closed parenthesis (**)**).

5.1.4   Operators

You may combine and manipulate functions as you would with other expressions. In general, operators should be entered *after* the right parenthesis (**)**) that terminates a function. For instance, to evaluate $\sin^2(x)$ (the square of the sine of x), you would enter **sin(x)^2**, but *not* "**sin^2(x)**" or "**sin (x^2)**"!

Step 103: Functions and Powers

> **sin(Pi/4)^2;**   *Determine* $\sin^2(\frac{\pi}{4}) = (\sin\frac{\pi}{4})^2$

$$\frac{1}{2}$$ $\sin(\frac{\pi}{4}) = \frac{1}{2}\sqrt{2}$. *Thus*, $(\frac{1}{2}\sqrt{2})^2 = (\frac{1}{4})(2) = \frac{1}{2}$.

Remember that Maple performs full evaluation and automatic simplification on expressions composed of functions, because the functions are also expressions.

5.1.5   Assignments

You may use assignments to simulate functions by entering the statement ***name:=expr***. Maple then replaces each instance of ***name*** with ***expr***:

Step 104: Simulated Function Assignment

> **y := 2*x+1;**   *Assign the expression* 2x+1 *to the variable y.*

$$y := 2x + 1$$ *So far, so good....*

> **x:=1: y;**   *Assign* 1 *to the variable x.*

$$3$$ *Maple replaced x with 1 inside 2x+1, which produces 3.*

However, you must never enter $f(x) = mx + b$ as "**f(x) := m*x+b**" to express $f(x)$:

Step 105: Incorrect Function Assignment

> **x:='x':**   *Unassign x.*

```
> f(x) := 2*x+1;
```
$$f(x) := 2x + 1$$
Try to assign a function.
The function appears OK but is really incorrect!

```
> f(1);
```
$$f(1)$$
Assign to the variable x the value 1.
Maple does not evaluate f(x) because
you used the incorrect notation.

Maple does permit valid functional notation *f(x)* syntax, which is explained in Section 5.6.

PRACTICE!

1. How many arguments does the expression **plot(x,x=0..10, title= "hello!")** have?
2. Enter the expression $\sin^2\theta + \cos^2\theta$ using Maple Notation.
3. Reduce the expression in the previous function with a call to the **simplify** function.
4. Simulate the equation $f(t) = \sin(t)$ with the name y and an assignment.

5.2   POLYNOMIALS

Polynomials are incredibly common functions because they help approximate intricate models. This section demonstrates many functions for operating on and manipulating polynomials. For an extensive overview of all of these functions, consult **Mathematics...
Algebra... Polynomials...** in the Help Browser.

5.2.1   Definition

Polynomials contain sums of *terms* with *integer* exponents:

$$(term)^0 + (term)^1 + (term)^2 + \ldots + (term)^n. \tag{5.1}$$

Each *term* represents virtually any name or constant, and not all terms and powers must be present. For example, both $x + 1$ and $x^2 + y + 30$ are polynomials. If you are unsure, you may confirm whether or not the expression is a polynomial:

Step 106: Check Polynomial Type

```
> x:='x':y:='y':
> type(x^2+y+30,polynom);
```
$$true$$
Clear x and y assignments.
Test whether or not $x^2 + y + 30$ is a polynomial.
$x^2 + y + 30$ is indeed a polynomial.

For more information on polynomial types, consult **?polynom**, **?content**, **?ratpoly**, **?type[polynom]**, and **?type[monomial]**.

5.2.2   Arithmetic

You may use Maple operators and functions on polynomials. Beware that sometimes Maple will not simplify the resulting expression, usually with multiplication and division! To demonstrate, try the following steps:

Step 107: Assign Polynomials

> **P1 := x^2+3*x+2:**   *Assign to* P1 *a polynomial* $x^2 + 3x + 2$.

> **P2 := x+4:**   *Assign to* P2 *a polynomial.* $x+4$.

Now, perform arithmetic operations:

Step 108: Polynomial Addition

> **P1+P2;**   *Add two polynomials.*
$$x^2 + 4x + 6$$ *Maple automatically simplifies common terms.*

Maple was able to add common terms. But, what about multiplication?

Step 109: Polynomial Multiplication

> **P1*P2;**   *Multiply two polynomials.*
$$(x^2 + 3x + 2)(x + 4)$$ *Chapter 7 discusses how to simplify these results.*

5.2.3   Expanding

Are you wondering why Maple did not produce $x^3 + 7x^2 + 14x + 8$ in Step 109? Maple does not prefer to multiply all terms because you might need automatic simplification for division operation in further evaluations. (See the practice problems that follow.) If you wish to force the multiplication, enter **expand(P1*P2)**:

Step 110: Expand Polynomials

> **expand(P1*P2);**   **expand** *performs all possible multiplication and addition operations.*
$$x^3 + 7x^2 + 14x + 8$$ $(x^2+3x+2)(x+4) = x^3+7x^2+14x+8$

According to **?expand**, **expand(expr)** distributes products over sums.

5.2.4   Factoring

You may "reverse" the effects of expand by using **factor(polynom)**. The *factors* of a polynomial are the smallest divisible polynomials whose product yields the polynomial. For the polynomials *P1* and *P2*, try to find their factors:

Step 111: Factor Polynomials

```
> factor(P1); factor(P2);                    Factor P1 and P2.
```
$$(x + 2)(x + 1)$$ $(x+2)(x+1) = x^2+3x+2 = P1$
$$(x + 4)$$ $(x+4)$ is the only factor of $x+4$.

Maple simply returns the polynomial as the sole factor when no factorization is possible. For related functions and more information, consult **?factor**, **?ifactor**, and **?roots**.

PRACTICE!

5. Determine whether or not the following statements produce a polynomial:
   `[> a:='a': x:=sin(a): poly:=x^2+2;`
6. Unassign x. Now, store $x^2 - 1$ in A. Store $x^2 + 3x + 2$ in B.
7. Evaluate $A^2 + AB$. Assign the result to $C1$.
8. Evaluate $\dfrac{A}{B}$. Assign the result to $C2$.
9. Enter **expand** and/or **factor** to simplify $C1$ and $C2$.

5.2.5 Division

Given two expressions a and b, you can divide a by b by specifying $a \div b$ or a/b. The division of a by b produces a *quotient* q and a *remainder* r such that

$$a = bq + r. \tag{5.2}$$

For instance, to prove that $5 \div 3$ produces a quotient of 2 and remainder of 1, let $a = 5$, $b = 3$, $q = 1$, and $r = 2$. Using Eq. 5.2, you will see that $5 = (3)(1) + 2$.

Now, suppose you want to divide $x^2 + 3x + 2$ by $x + 4$. Although you have two polynomials, Eq. 5.2 still holds. To demonstrate, you can divide two polynomials **a** by **b** in terms of **term** with either **quo(a, b, term, 'r')** or **rem(a, b, term, 'q')**. You may supply **'r'** or **'q'** as arguments to **quo** or **rem**, respectively, to automatically assign a remainder **r** or quotient **q** value:

Step 112: Divide Polynomials

```
> P1 := x^2+3*x+2:          Assign a polynomial  x^2+3x+2  to P1.
> P2 := x+4:                Assign a polynomial  x+4  to P2.
```

```
> q := quo(P1,P2,x,'r'):    Divide a by b such that  a = bq + r.
> q, r;                     Report the quotient q and remainder r.
```
$$x - 1, 6$$ $(x+4)(x-1)+6 = x^2+4x-x-4+6 = x^2+3x+2$

For more information, consult **?rem** or **?quo**. You will discover related functions in **?divide**, **?mod**, **?gcd**, **?lcm**, and **?evala**. For pure integer division, check out **?irem** or **?iquo**.

5.2.6 Root Finding

Factorable polynomials have **roots** that equate polynomials to zero when the roots are substituted back into the polynomial. For instance, $x^2 + 3x + 2$ factors into $(x + 2)$

$(x + 1)$ with roots $x = -2$ and $x = -1$, which both cause $x^2 + 3x + 2$ to become zero. Other polynomials have repeated factors, such as $x^2 + 2x + 1 = (x + 1)(x + 1)$, which has root -1 appearing twice. Thus, the root -1 has a *multiplicity* of 2.

Maple reports polynomial roots as a list of pairs in the form $[[r_1, m_1], [r_2, m_2], \ldots, [r_n, m_n]]$. Each $[r_i, m_i]$ pair is the ith root r with multiplicity m. For instance, to find the roots of the polynomial $P1 = x^2 + 3x + 2$, use **roots (polynom)**:

Step 113: Polynomial Roots

```
> x := 'x':                                          Unassign x.
> roots(x^2+3*x+2);                                  Find the roots of P1.
              [[-2, 1], [-1, 1]]             P1 has two roots, -2 and -1.
                                        -2 factors P1 only once. -1 factors P1 only once.
```

If you compare the results of Step 113 and **factor(x^2+3*x+2)**, you can verify that **roots** found each root of $x^2 + 3x + 2$. For more information, see **?roots**, **?root**, and **?realroot**.

PRACTICE!

10. Evaluate the quotient q and remainder r in Step 112 with **rem**.
11. Confirm that your quotient and remainder in the preceding problem are valid. Hint: Use **expand**.
12. What are the roots of $x^3 - 3x - 2$? Do any roots repeat? If so, how many times?

5.3   TRIGONOMETRY

Many equations rely on trigonometry to transform physical models into a variety of co-ordinate systems. (See **?coords**.) After all, nature knows no axes, however we might try to impose them! Trigonometry helps model problems throughout all branches of engineering and science. This section introduces basic trigonometric functions in Maple.

5.3.1   Warning! Use Radians for Angles

Many programs, including Maple, require angles to be entered in terms of **radians**. To perform the conversion, use the following ratios:

$$\frac{radians}{2\pi} = \frac{degrees}{360°}, \tag{5.3}$$

use **convert(angle,radians)** or **convert(angle,units,degrees,radians)**:

Step 114: Angle Conversion to Radians

```
> convert(45*degrees, radians);                      Enter 45° as 45*degrees.
              1
              ─π          Maple evaluates radians in terms of π when possible.
              4
```

TABLE 5.1   Trigonometric Functions

Function	Standard Math	Maple Notation	Description
sine	$\sin\theta$	`sin(theta)`	
cosine	$\cos\theta$	`cos(theta)`	
tangent	$\tan\theta = \dfrac{\sin\theta}{\cos\theta}$	`tan(theta)`	
cosecant	$\csc\theta = \dfrac{1}{\sin\theta}$	`csc(theta)`	
secant	$\sec\theta = \dfrac{1}{\cos\theta}$	`sec(theta)`	
cotangent	$\cot\theta = \dfrac{1}{\tan\theta}$	`cot(theta)`	

In the Description column (with a right triangle diagram labeled r, y, x, θ):

$$\sin\theta = \frac{y}{r} \qquad \csc\theta = \frac{r}{y}$$
$$\cos\theta = \frac{x}{r} \qquad \sec\theta = \frac{r}{x}$$
$$\tan\theta = \frac{y}{x} \qquad \cot\theta = \frac{x}{y}$$

Inverse trigonometric functions start with **arc**:

Standard Math: $\sin^{-1}\dfrac{y}{r} = \arcsin\dfrac{y}{r} = \theta$

Maple Notation: **arcsin(y/r)**

Remember to always specify π as **Pi** when entering angles with radians! Consult **?convert[degrees]**, **?convert[radians]**, and **?Units[angle]** for more information.

5.3.2   Trigonometric Functions

Table 5.1 summarizes common trigonometric functions. The following example demonstrates that Maple sometimes uses automatic simplification with trigonometric functions. Note the use of radians:

Step 115: Automatic Simplification and Trig

```
> sin(0), sin(Pi/2), sin(Pi);
```
Find $\sin(0), \sin(\frac{\pi}{2}), \sin(\pi)$

$0, 1, 0$ *Maple used automatic simplification to find the answers.*

Consult **?trig** for a full listing that includes hyperbolic functions. Inverse trigonometric functions are described in **?invtrig**.

PRACTICE!

13. Convert 120° to radians. Convert the result back to degrees.
14. Find the secant of 30°.
15. Find the tangent of $\dfrac{\pi}{2}$.
16. Assume the sine of an angle is 0.35. What is the angle in degrees?

5.4   POWERS AND ROOTS

This section reviews functions that are associated with powers and roots.

5.4.1   Exponentiation

Use the ***exponentiation*** operator ^ to raise an expression to a power. In the following example, try entering 123×10^{-2} without using scientific notation **e** or **E**:

Step 116: Exponentiation

> `> 123.*10^(-2);`   *Simulate scientific notation.* `123.*10**(-2)` *also works.*

$$1.230000000$$ *Yes, entering* `123.0E-2` *would be quicker.*

See also the following:

- `?arithop` and `?type[arithop]` for `^`.
- `?float` for scientific notation with `e` and `E`.

5.4.2 Roots

You have already used `sqrt(x)` for $\sqrt{x}$. In general, you can also find the *n*th **root** of *x* with exponentiation and fractional powers:

$$\sqrt[n]{x} = x^{\frac{1}{n}}. \tag{5.4}$$

For instance, try finding the cube root of 8:

Step 117: Roots

> `> 8^(1/3);`   *Find the cube root of 8.*

$$8^{\frac{1}{3}}$$ *Maple keeps the result in exact form.*

Maple wi ll not automatically simplify because in many cases the root is a float. To force evaluation, use floats or try `simplify(expr)`:

Step 118: Find Numerical Roots

> `> simplify(8^(1/3);`   *Try to simplify* $\sqrt[3]{8}$.

$$2$$ *Use* `evalf` *to produce a float.*

You may also enter `root(expr,integer)`:

Step 119: Use `root` to Find Roots

> `> root(8,3);`   *Find the cube root of 8.*

$$2$$ $2^3 = 8.$

`root` actually finds the ***principal root***, as explained in `?root`. Sometimes a principal root yields complex results:

Step 120: Generate Complex Roots

> `> simplify((-1)^(1/3));`   *Find the cube root of* −1.

$$\frac{1}{2} + \frac{1}{2}I\sqrt{3}$$ *What happened to the real root,* −1 ? $(-1)^3 = -1!$

The next section demonstrates how to find real roots when the principal root is complex. See also **?sqrt** and **?roots** for related functions.

5.4.3   Real Roots

If Maple does not generate a real root and you think one exists, try a function with a rather odd name called **surd**. Just as you would use **root**, enter **surd(expr, n)**. When n is odd, then

$$\operatorname{surd}(x, n) = \begin{cases} x^{1/n} & x \geq 0 \\ -(-x)^{1/n} & x < 0 \end{cases}. \tag{5.5}$$

These equations can generate real roots, especially for odd roots of negative numbers. For instance, find the real root of $(-1)^{1/3}$:

Step 121: Real Roots

> **surd(-1,3);**   *Find the non-complex cube root of* $-1 = (-1)^{1/3}$.

$$-1$$ *Maple found* $(-1)^{1/3} = -1$.

When no real root exists, **surd** returns a complex root. To generate real roots without having to worry about complex results and weird-sounding functions, use the **RealDomain** package as follows:

Step 122: Using **RealDomain** to Avoid Complex Numbers

> **with(RealDomain):**   *Enforce real roots and no complex-number results.*

Ignore Maple's warning message about redefined functions.

> **(-1)^(1/3);**   *Find the non-complex cube root of* $-1 = (-1)^{1/3}$.

$$-1$$ *Maple found* $(-1)^{1/3} = -1$.

When using **RealDomain**, Maple reports the extended numeric undefined for expressions that will only generate complex results:

Step 123: Generating Complex Numbers with **RealDomain**

> **sqrt(-1);**   *Without* **RealDomain**, *Maple would report 1.*

$$\textit{undefined}$$ *Maple cannot produce a complex number.*

For more information, consult **?surd**, **?arithop**, and **?RealDomain**.

5.4.4   Symbolic Roots

You might encounter another interesting problem when taking roots of symbolic expressions. For instance, try taking the square root of x^2. You will not obtain the obvious answer x:

Step 124: Problem with Symbolic Root

> `x:='x':`   *Unassign x.*

> `root(x^2,2);`   *Find $\sqrt{x^2}$.*

$$\sqrt{x^2}$$ *Maple does not know if x is positive or negative!*

Maple does not know if x is positive or negative, so Maple cannot evaluate the input any further. To convince Maple that you really want x, you must tell Maple that x^2 refers to a generic symbolic value:

Step 125: Symbolic Root

> `root(x^2,2,symbolic);`   *Find $\sqrt{x^2}$, assuming x is strictly symbolic.*

$$x$$ *Maple now finds the "right" answer.*

You could also instruct Maple to assume certain properties of variables. By entering **assume(x>=0)** Maple "knows" that x is positive for the entire session. You may also make a temporary assumption by entering **sqrt(x^2) assuming x>=0** all in one statement. The **RealDomain** package helps quite a bit. For information on applying many kinds of properties to variables for roots and other operations, investigate **?assume**, **?assuming**, and **?RealDomain**.

5.4.5   Logarithms

Consult Table 5.2 for a review of logarithms. A ***natural logarithm***, $\ln x$, employs the irrational base $e = 2.71828\ldots$ Logarithms of a general base b can be converted to ln form using the formula

$$\log_b y = \frac{\ln y}{\ln b}. \tag{5.6}$$

Maple usually expresses logarithms in terms of ln using the conversion in Eq. (5.6). For instance, find the base-10 log of 100:

Step 126: Logarithms

> `x:=log[10](100);`   *Evaluate $\log_{10} 100 = x$, where $10^x = 100$.*

$$x := \frac{\ln(100)}{\ln(10)}$$ *Maple prefers answers in terms of ln.*

> `simplify(x);`   *Also, try **evalf** for floating-point values.*

$$2$$ *$10^2 = 100$.*

For more information about logarithms, consult **?log** and **?ilog**.

TABLE 5.2   Logarithms

Function	Standard Math		Maple Notation
Logarithm of Base b	$b^x = y$	$\log_b y = x$	`log[b](y)`
Base 10 Logarithm	$10^x = y$	$\log_{10} y = x$	`log[10](y)`, `log10(y)`
Natural Logarithm	$e^x = y$	$\log_e y = \ln y = x$	`ln(y)`, `log(y)`

5.4.6   Exponential Function

To raise the constant e to a power x, do *not* enter "`e^x`"! Instead, enter the ***exponential function*** exp (x) which represents e^x. For instance, try the following inputs:

Step 127: Exponential Function

> `exp(x), exp(2), 3*exp(2*x);`   *Enter the sequence* e^x, e^2, $3e^{2x}$.

$$\mathbf{e}^x, \mathbf{e}^2, 3\mathbf{e}^{(2x)}$$

Maple outputs e as **e**.

Although Maple outputs **exp(*expr*)** as **e**^{expr}, you must never enter "**e**" or "**E**" to produce the exponential function! If you do not believe me, see **?E** for the proof. If you wish to use just e, enter **exp(1)** or select the "e" on the Symbol Palette. In both cases, Maple reports **e**. Also, do not confuse the exponentiation operator caret (**^**) with the exponential function **exp**. See **?exp** for more information.

PRACTICE!

> 17.   Evaluate $\sqrt[3]{-8}$ Find all real and complex roots.
> 18.   Find x such that $7^x = 163$. Show your answer as a float. Check the answer that Maple produces.
> 19.   Evaluate the exponential constant to five decimal places.
> 20.   Find $\ln(\exp(x))$. Based on your result, discuss the relationship between the functions ln and exp.

5.5   MISCELLANEOUS

Table 5.3 reviews common mathematical operations and functions you might encounter throughout your education and career in engineering and science. ***Procedural Maple functions***, like manipulation, evaluation, solving, plotting, and programming, are reviewed in later chapters.

PRACTICE!

> 21.   Evaluate $|-18|$, $|0|$, and $|18|$.
> 22.   Add the real and imaginary components of e^{ix}.
> 23.   Generate the sequence 1, 2, 4, 8 with **seq** or **$**. Assign the result to S.
> 24.   Add each element of S. Hint: Consult **?sum**.
> 25.   Multiply each element of S. Hint: Consult **?product**.

TABLE 5.3   Miscellaneous Functions and Operations

Functions	Standard Math	Maple Notation	Related Functions and Help
Absolute Value	$\lvert x \rvert$	`abs(x)`	`?abs, ?sign, ?signum, ?csgn`

Boolean	$x \wedge y$
	 $x \vee y$
	 Is $x \neq y$?
		`x and y`
		 `x or y`
		 `evalb(x <> y)`
			`?boolean, ?equation,`
			 `?evalb, ?logic`

Complex	$\mathrm{R}(x) + \mathfrak{I}(x)$ e^{ix}	`Re(x) + Im(x)` `exp(I*x)`	`?argument, ?conjugate,` `?csgn, ?evalc, ?polar`
Factorial	$x!$	`x!` `factorial(x)`	`?binomial, ?combinat, ?combstruct,` `?factorial, ?group`
Floats	$\sqrt{2} = 1.414\ldots$	`evalf(sqrt(2))`	`?evalf, ?float, ?fsolve,` `?numapprox, ?trunc`
Integer	$72 = (2)^3 (3)^2$	`ifactor(72)`	`?arith, ?ifactor, ?integer, ?trunc`
Inverse	f^{-1}	`f@@(-1)`	`?@, ?@@, ?invfunc`
List	$[x_1, x_2]$ $[f(x_1), f(x_2)]$	`[x[1],x[2]]` `map(f,[x[1],x[2]])`	`?list, ?member,` `?select, ?sort`
Piecewise	$f(x) = \begin{cases} e^x & 0 < x \\ 0 & \text{otherwise} \end{cases}$	`piecewise(0<x,exp(x))`	`?piecewise`
Product	$\prod_{i=1}^{n} x_i$	`product(x[i],i=1..n)`	`?mul, ?product`
Sequence	x_1, x_2, x_3	`seq(x[i],i=1..3)` `x[i] $ i=1..3`	`?seq, ?sequence, ?$`
Series	$\cos(x) = 1 - \frac{1}{2}x^2 + O(x^4)$	`series(cos(x),x=0,4)`	`?Order, ?powseries, ?series, ?taylor`
Set	$x \cap y$ $x \cup y$	`x intersect y` `x union y`	`?intersect, ?minus, ?set, ?union`
Summation	$\sum_{i=1}^{n} x_i$	`sum(x[i],i=1..n)`	`?add, ?sum`

5.6   FUNCTIONAL NOTATION

Entering assignments in the form of **y:=m*x+b** provides only a shortcut to simulate a function. This section discusses how entering functions in *functional notation* in the form of **f(x)** is more natural than using simulated functions.

5.6.1   Definition

Use the operator **->** to create your own functions that use functional notation in the form of **name(args)**. Assign the function **name** with the syntax **name:=args -> expr**:

- **name** defines the function name. Avoid using protected names.
- **args** are function arguments.
- **expr** is the actual expression of the function in terms of **args**.

You may use zero, one, or multiple arguments in **args**. Table 5.4 demonstrates the syntax for a variety of examples with different amounts of arguments. For further explanation, see the sections that follow and investigate **?->** and **?operators[example]**. To use a different method other than **->**, consult **?unapply**.

TABLE 5.4 Functional Notation

Arguments	Standard Math	Maple Notation
0	$f() = x$	`f := () -> x`
one	$f(x) = mx + b$	`f := x -> m*x + b`
multiple	$f(x, y) = x^2 + y^2$	`f := (x, y) -> x^2 + y^2`

5.6.2 Creating Functions in Functional Notation

For instance, to create a function of one variable, like $f(x) = x^2$, use the syntax **name:= var -> expr**:

Step 128: Functional Notation with One Variable

> `f :+ x -> x^2;`   *Assign* `f(x)` $= x^2$.

$$f := x \rightarrow x^2$$ *Maple now considers f as the functional form* `f(x)`.

Step 129: Show Values with Functional Notation

> `f(0), f(1), f(2), f(3);`   *Use* `f(x)` *to find four distinct values of x.*

$$0, 1, 4, 9$$ *Do not assign x to any expression. Instead, use* `f(x)`.

PRACTICE!

26. Create a function $dis(x) = x^3 + x^2 + x + 1$ using functional notation.
27. Find the value of $dis(x)$ at $x = -1$ and $x = 1$.
28. Plot $dis(x)$ on the interval $-1 \leq 0 \leq 1$.
29. Create a function $f(x, y, z) = x + y + z$ using functional notation.

5.7   APPLICATION

This section introduces aspects of hydraulics and fluid flow that apply various Maple functions and help you to use Maple's **Units** and **ScientificConstants** packages.

5.7.1 Background

Open-channel flow refers to fluid flow that is partially exposed to the atmosphere. Common examples include water flowing through rivers, canals, and sewers. Often, such flows need control to limit their height without completely blocking the flow. As shown in Figure 5.5, weirs resemble dams, except that the weirs permit overflow.

Sharp-crest weirs have flat, sharp surfaces like that of the triangular, or V-notch, weir that Figure 5.5 illustrates. Besides limiting flow, triangular weirs can aid flow-rate

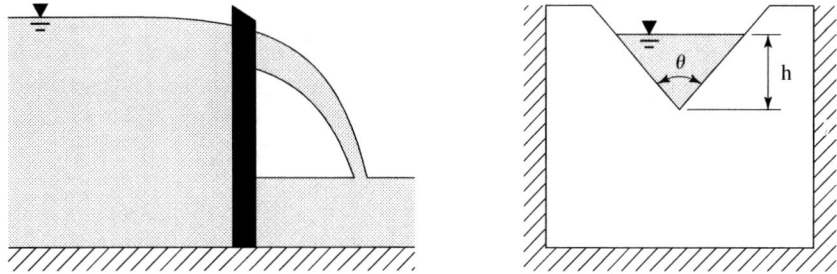

Figure 5.5   Triangular Weir

measurements, especially when rates greatly vary. The formula for flow Q (m³/s) over a triangular weir is

$$Q = C_d \left(\frac{8}{15}\right) \tan\left(\frac{\theta}{2}\right) \sqrt{2g} h^{2.5} \tag{5.7}$$

Q depends on the experimentally derived *discharge coefficient* C_d, the notch angle θ, and flow height h (m). Equation (5.7) uses the gravitational constant g, which is approximately equal to 9.81 m/s².

5.7.2   Problem

Given a triangular weir with notch angle 60° and flow height $h = 2.3$ m, determine the flow rate over the weir.

5.7.3   Methodology

I am using Maple's **Units** and **ScientificConstants** packages, so you should investigate Appendix C before continuing. Once you have become familiar with those packages, do the following steps:

Step 130: Weir—Initialize Maple

```
> restart:
```
Restart your Maple session.

```
> with(ScientificConstants):
```
Access Maple's built-in constants.

```
> with(Units[Natural]):
```
Use metric units in your calculations.

Ignore the long warning message that Maple issues.

Step 131: Weir—Restate and Separate

```
> my_g := evalf(Constant(g,units)):
```
Assign the gravitational constant.

$$my\_g := 9.80665 \left[\frac{m}{s^2}\right]$$

evalf *forces Maple to give you a numerical value.*

```
> h := 2.3*m:
```
Assign the flow height.

```
> theta := convert(60, units, degrees, radians):
```
Assign the notch angle.

Step 132:

For a "prettier" solution, use an indexed name to represent the coefficient C_d:

```
> C[d] := 0.59;
```
Express C_d as an indexed name to format the index d.

$$C_d := 0.59$$

Now your output looks like typeset mathematics.

Now, express the entire formula for flow rate Q:

Step 133: Weir—Model

```
> Q := C[d] * (8/15) *tan
(theta/2)*sqrt(2*g)*h^(5/2);
```
Model the flow rate.

$$Q := 3.7266998\sqrt{3}\left[\frac{m^3}{s}\right]$$

Units *kept track of all the units!*

For a more useful answer, force Maple to express Q in a numerical form:

Step 134: Weir—Solve and Report

```
> evalf(Q,3);
```
Attempt to reduce the numbers and units by simplification.

$$6.45\left[\frac{m^3}{s}\right]$$

Maple will not automatically reduce roots of variables.

KEY TERMS

argument	exponential function	exponentiation
factors	functional notation	Maple function
natural logarithm	nested function	parameter
polynomial	principal root	procedural Maple function
radian	root	

SUMMARY

- Functions take input, perform a task, and produce output.
- A function is a correspondence between the input (domain) and the output (range).
- You may nest Maple functions by supplying a function call as input to another function.
- You may combine functions with operators.
- Polynomials contain sums of terms with integer exponents.
- Trigonometric functions require radians (instead of degrees) for angles.
- You may find roots with **root** or the operator **^**.
- When using **root**, Maple finds the principal root, which may be complex.
- To find a real root, use **surd** or the **RealDomain** package.

- Maple expresses logarithms in terms of the natural log function **ln**.
- Maple expresses the exponential function with **exp**.
- To create a function **f** with the notation **f(x)**, use the **->** operator.

Problems

1. What is a function? How do you express functions in Maple?

2. Given $f(x) = 2x^3$, evaluate $f(-1)$, $f(0)$, and $f(1)$ using Maple using a simulated function.

3. For Problems 3a through 3c, let $P = x^2 + 6x + 7$ and $Q = x + 1$.

   (a) Evaluate $P + Q$ and $P - Q$.

   (b) Evaluate PQ, P^2Q, and $\dfrac{P}{Q}$. Distribute (multiply out) all products and sums.

   (c) Divide P by Q using both **rem** and **quo**. Show the quotient and remainder in both cases. Hint: For instance, enter **rem(P, Q,x,'q')** for **rem**.

   (d) Confirm your results in Problem 3c. Hint: Try both **evalb** and **expand**.

4. Factor the polynomial $x^4 - 2x^2 + 1$. How many different roots does the polynomial contain? How many times does each root factorize the polynomial? Hint: Try **factor** and **roots**.

5. Evaluate $\sqrt[3]{-72}$. Show both real and complex roots. Hints: Use **simplify** to clarify the results. Note that Maple will show fractional components.

6. Can you take the natural logarithm of a negative number? Demonstrate your answer with a plot of $\ln(x)$ on $-1 \le x \le 1$.

7. Solve the equation $y = x^b$ for x by taking the natural log of both sides of the equation. Using **solve**, solve the resulting equation for x.

8. Evaluate the following expressions:

   (a) $\sin^2(x)$

   (b) $\sin\left(\frac{\pi}{4}\right)$

   (c) $\sin^2\left(\frac{\pi}{4}\right)$

   (d) $\sqrt{\sin(17)}$ (produce both exact and decimal results)

   (e) $\tan(45°)$

   (f) $\log_{10}100$

   (g) $\ln 5.216$

   (h) $2.4^{-1.2}$

   (i) $\dfrac{1}{e}$ (produce both exact and decimal results)

   (j) πe^2 (produce both exact and decimal results)

9. Create a function that finds a trapezoid's area

$$\text{trap}(b_1, b_2, h) = h \times \left(\frac{b_1 + b_2}{2}\right)$$

using functional notation. Evaluate trap(1, 2, 3) to test your function.

10.   Snow blowing over large unblocked distances called *fetch* contributes to accumulating snow drifts. Given the relationship between snow transport capacity $\frac{Q_t}{Q_{inf}}$ and fetch F (m)

$$\frac{Q_t}{Q_{inf}} = \left(1 - 0.14^{\frac{F}{3000}} \right)$$

Compute F assuming $\frac{Q_t}{Q_{inf}} = 0.8$. Does transport capacity increase or decrease as fetch increases? Hints: Rearrange the equation with logarithms on both sides. Also, consider entering **assume(F>0)**.

6

Manipulating Expressions

6.1　INTRODUCTION

Expressions have many equivalent forms. Accordingly, Maple might not always generate the answers that you expect. This chapter introduces ***expression manipulation***, which is a process to change an expression into an equivalent form. You might have already used functions that manipulate expressions, such as **simplify**, **factor**, and **expand**. Consult Mathematics...Algebra... in the Help Browser for a large collection of functions that help you to manipulate expressions. Also, try right-clicking Maple output: Maple will give you suggestions on a variety of functions to help you "tinker" with the program.

OBJECTIVES

After reading this chapter, you should be able to

- Develop skills for manipulating Maple expressions
- Convert expressions into other forms
- Simplify and reduce expressions into smaller forms
- Expand expressions into larger forms
- Sort and extract expressions

Automotive engineers strive to produce efficient, affordable, and reliable vehicles. The DIATA (Direct Injection Aluminum Through-Bolt Assembly) engine powers Ford's Hybrid Electric Vehicle called the *P2000*. With this engine, the five-passenger P2000 vehicle can achieve over 60 miles per gallon (mpg) with performance and comfort comparable to that of conventional cars. Courtesy of Ford Motor Company.

PROFESSIONAL SUCCESS: WRITE WELL, NOT "GOOD!"

Strive for clarity! Treat your writing as you would your calculations. Study these writing tips that I have adapted from a popular file floating around the Internet, which is usually called *Writing Tips*:

- Prepositions are not words to end sentences with.
- Avoid clichés like the plague. Besides, they are old hat.
- Pronouns confuse readers. They are uncertain, so avoid them, unlike this.
- Don't use contractions that aren't necessary.
- It was, perhaps, decided that sentences, like this, might possibly avoid responsibility.
- Avoid ampersands & abbreviations, etc.
- Parenthetical remarks (however relevant) are not necessary.
- Profanity sucks.
- "Alternative" is the better alternative between the alternatives, "alternative" and "alternate".
- Be more or less specific.
- Exaggeration is a billion times better than understatement.
- One-word sentences? Eliminate!
- On the other hand, excessively long sentences detract from readability and clarity by not only providing a barrage of detailed concepts, but by boring the reader as well, hence obfuscating pertinent ideas originally intended for incisive communication as this verbose sentence, which blathers on excessively to make a relatively simple point.
- The passive voice is to be avoided.
- Rhetorical questions? Who needs them?
- It's you're roll too reed four homophones. Sew than, due ewe sea there affects hear?
- Eschew sesquipedalianism.

6.2   CONVERSION

Maple expressions have a *type*, which is the dominant form, of an expression. Review **?type** for more information. This section reviews the *conversion* of expressions into other types, using **convert(expr, form)** to cast, or convert, expressions into other types. Other sections demonstrate the use of convert for unit and angle conversions. For example, did you know that $e^{i\pi}$ has a trigonometric form?

Step 135: Converting Expressions

```
> restart:
```
Clear all assignments.

```
> convert(exp(I*x),trig);
```
Convert e^{ix} into sin's and cos's.

$$\cos(x) + I \sin(x)$$
Yes, this expression is equivalent to e^{ix}.

The **convert** function also helps to swap between set and list types:

Step 136: Converting Expressions

```
> convert({a,b,c},list);
```
Convert the set $\{a, b, c\}$ into a list.

$$[a, b, c]$$
Square brackets indicate a list.

Consult **?convert** to see the types that **convert** can handle. Enter **?convert[type]** for help on converting to and from **type.**

PRACTICE!

> 1. Maple uses **RootOf** as a place holder for roots of expressions. Enter **RootOf(x^2+1=0)**. Now, convert the entire expression into Maple's radical form.
> 2. Convert the integer 438 into a float.

6.3   SIMPLIFICATION

Simplification usually implies the reduction of an expression into a more condensed, simpler form. Regardless of Maple's automatic simplification capabilities, Maple often needs additional coaxing to simplify expressions further. This section demonstrates functions for simplifying expressions. See also Mathematics...Algebra...Expression Manipulation...Simplifying... in the Help Browser.

6.3.1   Simplify

Generally, you should first attempt to use the **simplify** function when you want to simplify an expression. In general, enter **simplify(expr)** to simplify *expr*:

Step 137: Simplification

```
> A := (x+1)/(x^2-1);
```
Simplify $\dfrac{x + 1}{x^2 - 1}$ which equals $\dfrac{x + 1}{(x + 1)(x - 1)}$.

$$A := \frac{x + 1}{x^2 - 1}$$
Automatic simplification cannot find the common factor x+1.

> **simplify(A);**

$$\frac{1}{x-1}$$

Simplify $\dfrac{x+1}{x^2-1}$.

simplify *removed the common factor.*

However, not all simplifications produce the "smallest-looking" expressions:

Step 138: Simplest ≠ Smallest!

> **simplify(sin(x)^2);**

$$1 - \cos(x)^2$$

Simplify $\sin^2 x$.

Maple prefers cosine to sine.

Nor do all simplification statements actually appear to work:

Step 139: Simplification Might Fail!

> **simplify(sin(2*x)/cos(2*x));**

$$\frac{\sin(2x)}{\cos(2x)}$$

Does Maple produce $\tan 2x$?

Maple does not think $\tan 2x$ *is simpler!*

So, what should you do when the "pure" version of **simplify** does not seem to work? Try **simplify(expr, form)**, where **form** includes a variety of labels listed on **?simplify**. For instance, Maple does not automatically simplify radicals when the term is symbolic because the variable might represent a complex value. By using the symbolic form, you can force Maple to treat the term as strictly symbolic:

Step 140: Simplification with a **symbolic** Form

> **sqrt(x^2);**

$$\sqrt{x^2}$$

Find $\sqrt{x^2}$.

Maple refuses because x might be negative.

> **simplify(sqrt(x^2),symbolic);**

$$x$$

Require Maple only to consider the symbolic nature of x.

Now, Maple can find $\sqrt{x^2} = x$.

You will likely find the **size**, **power**, **trig**, and **power** forms quite useful. See **?simplify** for a full list and links to help on each form.

6.3.2   Combine

When your expressions involve sums, products, and powers, enter **combine(expr)** to simplify **expr**. The combine function's built-in transformation rules help reduce the size of an expression. For instance, Maple knows many trigonometric identities:

Step 141: Simplifying Expressions

> **combine(sin(x)*cos(y)-cos(x)*sin(y));**

$$\sin(x-y)$$

Does Maple know this trigonometric identity?

Maple might also report −sin(−x+y).

Enter **combine(*expr*, *form*)** for specific transformations where ***form*** is a name or list of names:

Step 142: Combining Expressions Using a Specified Form

> **A:=(x^a)*(x^b): combine(A);**   *Attempt to reduce* $(x^a)(x^b)$.

$$x^a x^b$$ **combine(A)** *does not suffice.*

> **combine(A,power);**   *Simplify* $(x^a)(x^b)$ *by combining the powers.*

$$x^{(a+b)}$$ *Supplying the* **power** *rule helps* **combine**.

Investigate **?combine** for its possible transformations. Also, consult **?combine[*form*]** for rules concerning each ***form***.

6.3.3 Normal

Rational numbers express numbers as integer ratios. Similarly, rational functions are ratios of polynomials. The **normal** function seeks rational functions in the ratio $\frac{numerator}{denominator}$ without common factors, which is also called **factored normal form**. For instance, you can eliminate common factors with **normal(*expr*)**:

Step 143: Rational Expressions

> **normal((x+1)/(x^2-1));**   *Remove the common factor* $x-1$.

$$\frac{1}{x - 1}$$ *Try* **simplify** *as well.*

In general, use **normal** for expressions that are in terms of sums and products. See **?normal** for more information.

6.3.4 Radicals

A ***radical expression*** (or simply *radical*) in Maple has the syntax **expr^(*fraction*)**, as discussed in **?type[radical]**. Maple provides many functions for simplifying and converting radicals. For instance, try **rationalize** to remove radicals from an expression's denominator:

Step 144: Radical Expressions

> **A:=1/(1+2^(1/3));**   *Assign an expression that contains the radical* $2^{\frac{1}{3}}$.

$$A := \frac{1}{1 + 2^{(\frac{1}{3})}}$$ *Maple automatically keeps the radical in the denominator.*

> **rationalize(A);**   *Remove radicals from the denominator.*

$$\frac{1}{3} - \frac{1}{3} 2^{(\frac{1}{3})} + \frac{1}{3} 2^{(\frac{2}{3})}$$ *All radicals now appear in the numerator.*

Maple provides many related functions: **?radsimp**, **?radnormal**, **?rationalize**, **?simplify[radical]**, **?combine[radical]**, and **?convert[radical]**.

6.3.5 Square Roots

Square roots are radicals that have the exponent 1/2, such as $x^{\frac{1}{2}} = \sqrt{x}$, as discussed in **?type[sqrt]**. Many simplification and conversion functions for radicals also handle square roots. For instance, try reducing a square root with **simplify (expr, sqrt)**:

Step 145: Square Root Expressions

```
> 12^(1/2);
```
$$\sqrt{12}$$

Try to find $\sqrt{12}$.

Maple does not automatically simplify this expression.

```
> simplify(%,sqrt);
```
$$2\sqrt{3}$$

Simplify the square root expression.

Maple evaluated $\sqrt{12}$ as $\sqrt{4}\sqrt{3}$ and reduced the expression.

The square root function **sqrt** embodies many of the same simplifications. For more information, investigate the same functions in Section 6.3.4 and **?simplify[sqrt]**, **?convert[sqrfree]**, and **?convert[RootOf]**.

PRACTICE!

3. Convert $\sin(x)$ into an expression involving e^x. Simplify the resulting expression such that the original expression reappears.
4. Demonstrate the trigonometric identity $\sin^2 x + \cos^2 x = 1$. Try both **combine** and **simplify**.
5. Does **combine** reduce the expression $\sin x \sin y$? Why or why not?
6. Simplify the expression $x(x + 1) - x^2$.
7. Express $x + \dfrac{y}{x}$ in factored normal form.
8. Simplify the expression $\dfrac{x^2 - 1}{\sqrt{x + 1}}$ by eliminating the radical in the denominator.

6.4 EXPANSION

Expression *expansion* usually reverses the effects of **combine** and **factor**. You may have used **expand** to distribute polynomial factors. Actually, **expand(expr)** distributes products of sums inside expressions:

Step 146: Expanding Expressions

```
> expand( (x+1)*(x-1) );
```
$$x^2 - 1$$

expand *multiplies factored polynomials.*

Sometimes **expand** *actually simplifies an expression.*

The **expand** function also understands many other functions:

Step 147: Expanding Expressions

> `expand( cos(x-y) );`   *expand will often generate identities used to simplify expressions.*

$$\cos(x)\cos(y) + \sin(x)\sin(y)$$ *Enter* **combine** *to reverse this expansion.*

Investigate **normal(*expr*,expanded)** for rational expressions. Consult **?expandon**, **?expandoff**, and **?factor** for more information.

PRACTICE!

9. How does Maple evaluate **expand(factor(x^2+3*x+2))**?
10. Simplify the resulting expression of **combine(sin(x)*sin(y))**.
11. Try both **expand** and **normal** to expand the expression $\dfrac{x + 2}{(x + 3)^3}$.

6.5   EXTRACTION AND SORTING

Structural manipulations of expressions include component separation, term rearrangement, and operand extraction as introduced in this section.

6.5.1   Rational Expressions

Enter **numer** and **denom** to extract the numerator and denominator, respectively, of a rational expression:

Step 148: Extract Numerator and Denominator

> `A := (a+b)/(c+d):`   *Assign a rational expression.*

$$A := \frac{a + b}{c + d}$$ *Maple displays your fraction.*

> `numer(A), denom(A);`   *Extract the numerator and denominator of A.*

$$a + b, c + d$$ *Maple produces the numerator and denominator.*

6.5.2   Polynomials

Enter **collect(*expr*,*term*)** to rearrange coefficients around *term*:

Step 149: Collecting Like Terms

> `P := x*y-(x^2+1)*y;`   *Assign a polynomial to P.*

$$P := xy - (x^2 + 1)y$$ *This is the uncollected polynomial.*

> `P2 := collect(P,y);`   *Rearrange coefficients of P around y.*

$$P2 := (x - x^2 - 1)y$$ **collect** *rearranged all coefficients.*

Now, enter **sort(*expr*)** to sort the coefficients in order of powers:

Step 150: Sort Terms

> **sort(P2);**   *Sort the coefficients and terms of P2.*

$$(-x^2 + x - 1)y$$

Maple sorts the terms from highest to lowest power.

You can also rearrange lists and sequences with **sort**. Investigate related commands inside Mathematics...Algebra...Polynomials... and in **?coeff**, **?lcoeff**, **?degree**, and **?split**.

6.5.3   Operands

Recall how expression trees connect subexpressions composed of Maple's language elements. The operand function **op** extracts these elements from an expression tree:

- **op(*expr*)** shows all operands contained in the expression's surface type.
- **op(0,*expr*)** determines the expression's surface type.
- **op(i,*expr*)** extracts the ith operand from **_expr_**'s surface type.

Now, experiment finding operands using a simple expression:

Step 151: Operands

> **a:='a': b:='b': expr:=a+b:**   *Assign a+b to expr.*

> **op(expr);**   *Show an expression's surface-type operands.*

$$a, b$$

a+b has two operands: a and b.

> **op(0,expr);**   *Evaluate the expression's surface type.*

$$+$$

Addition (+) connects the two operands a and b.

Also, consult **?nops**, **?subsop**, **?applyop**, and **?map**.

6.5.4   LHS and RHS

Many types of expressions, such as equations, inequalities, and ranges, employ the syntax **_LHS_** *operator* **_RHS_**, where:

- LHS represents the left-hand side of the expression.
- RHS represents the right-hand side of the expression.

For instance, in the equation $y = mx + b$, LHS $= y$ and RHS $= mx + b$. The relational operator, equals sign ($=$), connects both sides of the equation. You can extract the LHS and RHS of **_expr_** with **lhs(*expr*)** and **rhs(*expr*)**, respectively:

Step 152: LHS and RHS

> **restart:**   *Clear all variable assignments.*

> **EQN := y=m*x+b;**   *Assign the equation y = mx+b to EQN.*

$$EQN := y = mx + b$$ *The variable EQN stores the entire equation y = mx+b.*

> **lhs(EQN);**   *Extract the LHS of EQN.*

$$y$$ *The LHS of EQN is not the name EQN.*

> **rhs(EQN);**   *Extract the RHS of EQN.*

$$mx + b$$ *The LHS of EQN is not the name EQN.*

PRACTICE!

12. Arrange the polynomial $xy - (x^2 + 1)y$ about x. Now, sort the new polynomial.

13. Assign $1 + \sin(x + y)$ to A. Determine the surface type and operands of A.

14. Assign the equation $p = ku$ to *Spring*. Next, show the value assigned to *Spring*.

15. Evaluate **lhs(2>1)** and **rhs(2>1)** and inspect the output. Why does Maple reverse the relation $2 > 1$?

6.6   APPLICATION: TORSIONAL ANALYSIS

This section demonstrates Maple's expression-manipulation functions with an example from structural mechanics.

6.6.1   Background

Consider an I-beam (pronounced "eye" beam) that is loaded transversely with a uniform torque, which is a bending moment that twists the member, as illustrated in Figure 6.1. As expected, the member rotates about the z-axis at an angle ϕ. Most noncircular structural members, like this I-beam, also warp when twisted. That is, the cross section does not remain plane along the z-axis. You can simulate this warping effect by twisting a soft rectangular object like an eraser or a long cardboard box.

When determining how the member resists the applied torque, you can calculate the angle of twist ϕ with the formula

$$\phi = C_1 + C_2 z + C_3 \cosh\frac{z}{a} + C_4 \sinh\frac{z}{a} - \frac{a^2 m z^2}{2EC_w} \tag{6.1}$$

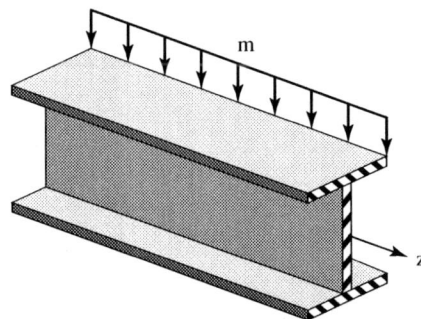

I-Beam with Torsional Load   Twisted Cross Section   Warped Section

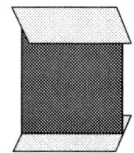

Figure 6.1   Warping Torsion

where the coefficients a, C_1, C_2, C_3, C_4, C_w, and E represent structural parameters and boundary conditions. Equation (6.1) is derived from the differential equation

$$\frac{1}{a^2}\frac{d^2\phi}{dz^2} - \frac{d^4\phi}{dz^4} = -\frac{m}{EC_w}, \tag{6.2}$$

which uses structural parameters C_w and E along with derivatives $d^2\phi/dz^2$ and $d^4\phi/dz^4$. Don't worry about the calculus—the functions are supplied for you. Instead, concentrate on how the expressions are manipulated.

6.6.2 Problem

When Maple solves Eq. (6.2) Maple's derivation does not look like Eq. (6.1). Manipulate Maple's result to prove that Eq. (6.1) is indeed a valid solution.

6.6.3 Methodology

First, blindly accept the input and output that follow to prove that Maple really does not provide Eq. (6.1) as the solution to Eq. (6.2). After taking a few calculus courses, these equations will not seem so difficult.

Step 153: Torsional Analysis—Initialize Maple

```
> restart:
```
Restart your Maple session.

Step 154: Torsional Analysis—Restate

```
> DE := (1/a^2)*diff(phi(z),z$2)
- diff(phi(z),z$4) = (-m)/(E*C[w]);
```
Enter Eq. 6.2.

$$DE := \frac{\frac{\partial^2}{\partial z^2}\phi(z)}{a^2} - \left(\frac{\partial^4}{\partial z^4}\phi(z)\right) = -\frac{m}{EC_w}$$

This monstrosity is a differential equation.

```
> phi1 := dsolve(DE,phi(z));
```
Don't worry for now. Just replicate what you see.

$$\phi1 := \phi(z) = -\frac{1}{2}\frac{a^2mz^2}{EC_w} + \_C1 + \_C2z + \_C3e^{\left(\frac{z}{a}\right)} + \_C4e^{\left(-\frac{z}{a}\right)}$$

Here is Maple's solution.

Now, use structural-manipulation functions to express $\phi1$ in terms of hyperbolic functions. For now, ignore the $\phi(z)$ portion on the LHS, and convert only the $\phi1$ portion on the RHS, instead. To convert the exponential functions, use a trigonometric conversion:

Step 155: Torsional Analysis—Convert

```
> phi2 := convert(rhs(phi1),trig);
```
sinh and cosh resemble sin and cos.

$$\phi2 := -\frac{1}{2}\frac{a^2mz^2}{EC_w} + \_C1 + \_C2z + \_C3\left(\cosh\left(\frac{z}{a}\right) + \sinh\left(\frac{z}{a}\right)\right)$$

$$+ \_C4\left(\left(\cosh\left(\frac{z}{a}\right) - \sinh\left(\frac{z}{a}\right)\right)\right)$$

Throughout this example, Maple might organize your output terms differently, but it will still produce equivalent equations. Now, reorganize $\phi2$ by using **collect** for each function:

Step 156: Torsional Analysis—Collect

> **phi3 := collect(phi2,cosh);**   *Collect terms in $\phi3$ corresponding to* **cosh**.

$$\phi3 := (\_C3 + \_C4)\cosh\left(\frac{z}{a}\right) - \frac{1}{2}\frac{a^2mz^2}{EC_w}$$

$$+ \_C1 + \_C2z + \_C3\sinh\left(\frac{z}{a}\right) - \_C4\sinh\left(\frac{z}{a}\right)$$

> **phi4 := collect(phi3,sinh);**   *Collect terms in $\phi4$ corresponding to* **sinh**.

$$\phi4 := (\_C3 - \_C4)\sinh\left(\frac{z}{a}\right) + (\_C3 + \_C4)$$

$$\cosh\left(\frac{z}{a}\right) - \frac{1}{2}\frac{a^2mz^2}{EC_w} + \_C1 + \_C2z$$

6.6.4   Solution

The equation for $\phi4$ is essentially the same as Eq. (6.1). If you prefer, see other sections for the function **subs**:

Step 157: Torsional Analysis—Substitute

> **phi := subs({_C1=C1,_C2=C2,(_C3+_C4)=C3, (_C3-_C4)=C4},phi4);**

$$\phi := C3\cosh\left(\frac{z}{a}\right) + C4\sinh\left(\frac{z}{a}\right) - \frac{1}{2}\frac{a^2mz^2}{EC_w} + C1 + C2z$$

You might also wish to further rearrange terms with **sort**.

KEY TERMS

conversion	expansion	expression manipulation
factored normal form	LHS	radical expression
rational function	RHS	simplification
structural manipulation		

SUMMARY

- To simplify an expression, start with **simplify** and its options.
- To convert an expression from one type to another, use **convert**.
- To merge subexpressions and produce "smaller" expressions, use **combine**.
- To reverse the effects of **combine**, use **expand**.
- To reorganize an expression, use **sort**, **collect**, and related functions.
- To extract operands, use **op**, **lhs**, and **rhs**.

Problems

1. Distinguish between conversion and simplification.
2. Under what conditions should you attempt to use the **normal** function?
3. Confirm the following identities using Maple:

(a) $\sin^2 x + \cos^2 x = 1$

(b) $1 + \tan^2 x = \sec^2 x$

(c) $1 + \cot^2 x = \csc^2 x$

(d) $\sin(x + y) = \sin x \cos y + \cos x \sin y$

(e) $\cos(x + y) = \cos x \cos y - \sin x \sin y$

(f) $\tan(x + y) = \dfrac{\tan x + \tan y}{1 - \tan x \tan y}$

(g) $\sin 2x = 2 \sin x \cos x$

(h) $\cos 2x = 2 \cos^2 x - 1$

(i) $\cos 2x = \cos^2 x - \sin^2 x$

(j) $\sin^2 x = \dfrac{1 - \cos 2x}{2}$

(k) $\cos^2 x = \dfrac{1 + \cos 2x}{2}$

(l) $\sin x + \sin y = 2 \sin\left(\dfrac{x + y}{2}\right)\cos\left(\dfrac{x - y}{2}\right)$

(m) $\cos x + \cos y = 2 \cos\left(\dfrac{x + y}{2}\right)\cos\left(\dfrac{x - y}{2}\right)$

(n) $\sin x \sin y = -\dfrac{1}{2}(\cos(x + y) - \cos(x - y))$

(o) $\cos x \cos y = \dfrac{1}{2}(\cos(x + y) + \cos(x - y))$

(p) $\sin x \cos y = \dfrac{1}{2}(\sin(x + y) + \sin(x - y))$

(q) $x^a x^b = x^{(ab)}$

(r) $\dfrac{x^a}{x^b} = x^{(a-b)}$

(s) $(x^a)^b = x^{ab}$ (assume real values)

(t) $(xy)^a = x^a y^a$ (assume real values)

(u) $\log_b xy = \log_b x + \log_b y$

(v) $\log_b \dfrac{x}{y} = \log_b x - \log_b y$

(w) $\log_b x^a = a \log_b x$

(x) $\sinh x = \dfrac{e^x - e^{-x}}{2}$

(y) $\cosh x = \dfrac{e^x + e^{-x}}{2}$

(z) $\tanh x = \dfrac{e^x - e^{-x}}{e^x + e^{-x}}$

(aa) $\cosh^2 x - \sinh^2 x = 1$

(ab) $\cosh 2x = \cosh^2 x + \sinh x$

(ac) $e^{ix} = \cos x + i \sin x$

(ad) $\sin x = \dfrac{e^{ix} - e^{-ix}}{2i}$

(ae) $\cos x = \dfrac{e^{ix} + e^{-ix}}{2i}$

(af) $\tan x = -i\left(\dfrac{e^{ix} - e^{-ix}}{e^{ix} + e^{-ix}}\right)$

4. Express $\dfrac{\sin 2x}{\cos 2x}$ in terms of a tangent function $\tan 2x$ using Maple. Hint: **simplify** will not work. Try another function from this chapter.

5. The equation $e^{i\pi} + 1 = 0$ combines five of the most famous symbols in mathematics. Confirm this relationship using Maple.

6. Confirm that $(x + y)^3 = x^3 + 3x^2 y + 3xy^2 + y^3$.

7. Confirm that $x^3 - y^3 = (x - y)(x^2 + xy + y^2)$.

8. Express $\dfrac{1}{1 - \sqrt{5}}$ such that no radical appears in the denominator.

9. Given the Maple session portion

```
> op(A);
> op(0,A);
```
$$ax^2,\, x,\, 2$$
$$+$$

determine the expression assigned to A. Indicate the first, second, and third operands using the **op** function.

10. A force applied to a body creates *stress*, which is a force distribution applied on a surface. Stress usually is expressed in units of force per unit area. For any surface, stress can be resolved into tangential (*shear*) or perpendicular (*normal*) components. However, rotating the perspective—that is, the coordinate system that defines the body's orientation—can produce a configuration that causes only normal stresses, as shown in Figure P6.2:

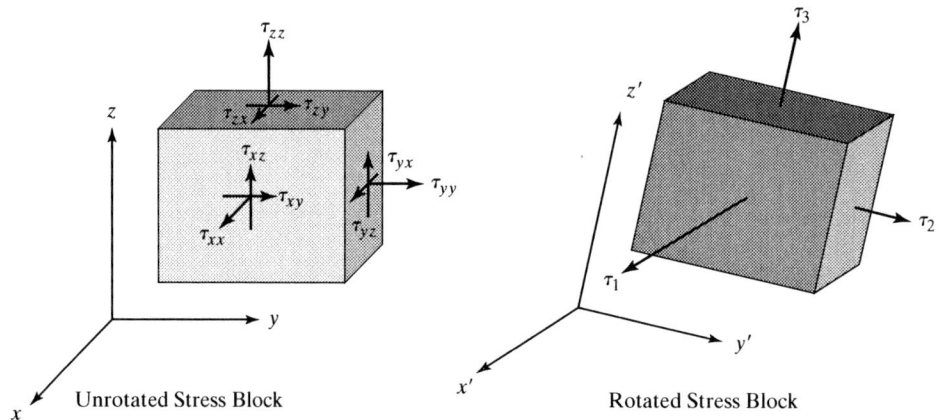

Figure P6.2 Stress Block

For a general orientation, each surface of the unrotated block in Cartesian (x, y, z) coordinates yields two shear-stress components and one normal-stress component per face. A certain rotated axis can produce only normal stresses called *principle stresses* τ_1, τ_2, and τ_3. You can find principle stresses from the formula

$$\tau^3 - A\tau^2 + B\tau + C = 0 \tag{6.3}$$

where

$$A = \tau_{xx} + \tau_{yy} + \tau_{zz} \tag{6.4}$$

$$B = \tau_{xx}\tau_{yy} + \tau_{yy}\tau_{zz} + \tau_{zz}\tau_{xx} - \tau_{xy}^2 - \tau_{yz}^2 - \tau_{zx}^2 \tag{6.5}$$

$$C = \tau_{xx}\tau_{yy}\tau_{zz} + 2\tau_{xy}\tau_{yz}\tau_{xz} - \tau_{xy}^2\tau_{zz} - \tau_{yz}^2\tau_{xx} - \tau_{zx}^2\tau_{yy} \tag{6.6}$$

Assume that a body has internal stresses $\tau_{xx} = 100$ kPa, $\tau_{yy} = 100$ kPa, $\tau_{zz} = 25$ kPa, $\tau_{xy} = \tau_{yx} = 10$ kPa, $\tau_{yz} = \tau_{zy} = 0$ kPa, and $\tau_{xz} = \tau_{zx} = 0$ kPa.

(a) Assign variables to the values for each stress.

(b) Evaluate the constants A, B, and C.

(c) Note that all shear stresses associated with the z-axis are zero. You can immediately determine τ_3. Hint: It's τ_{zz}.

(d) Solve for the three principle stresses, using Maple's polynomial-division capabilities. Hints: You already know one root already: $\tau = \tau_{zz} = 25$ kPa. Hence, now you need to divide Eq. (6.3) by the factor $\tau - 25$.

7

Graphics

7.1   INTRODUCTION

Formulas model physical processes, and in turn, the graphical representations of models help you visualize qualitative physical behaviors. Maple graphics primarily include **plots**, which are graphs of points, expressions, and equations. This chapter delves into Maple's graphical tools.

OBJECTIVES

After reading this chapter, you should be able to

- Locate plotting functions and packages
- Manipulate and customize plots
- Plot two-dimensional expressions
- Plot three-dimensional expressions
- Display multiple plots
- Plot text and legends
- Animate plots

7.1.1 Tutorials and Sources

Maple's plotting capabilities are very comprehensive, and Maple provides various ways to access information about these tools:

- *Help:* How To... Plot... Overview and Graphics...
- *Menus:* Reference... menus... Overview
- *Context Bar:* **?context2dplot** and **?context3dplot**
- *Context Menu:* **?contextmenu**
- *Examples:* **Mathematical Visualization** inside **?examples[index]**
- *Packages:* Graphics... Packages...
- *Interactive help:* **?plots[interactive]**, **?smartplot**

7.1.2 Library Packages

Consult **?DEtools, ?plottools, ?plots, ?stats, ?geometry**, and **?geom3d** on a large variety of graphics-function libraries. Appendix B explains how to access the library packages. In this chapter, you will primarily use **with(package)**.

7.1.3 Categories

Generic functions for plotting images include the following:

- **plot**: Create two-dimensional plots using command line instructions (Section 7.2).
- **smartplot**: Create plots using the GUI (**?smartplot**).
- **display**: Display plots assigned to names (Section 7.3).
- **animate**: Animate plots of functions and equations (Section 7.5.7).
- **draw**: Show images created using the **geometry** library package (Section 7.5.6).

Many plotting functions also support three-dimensional versions. To call these functions, append the label **3d** to the function name, as in **plot** becoming **plot3d**.

PROFESSIONAL SUCCESS: TEAMWORK

Learning often is difficult. Avoid the "lone wolf" mentality by seeking help from professors and teaching assistants when you are in need. Also, your classmates often suffer through the same problems as you, so you should consider working with them. Perhaps another student knows a trick you missed. Trading help also helps you learn. Better yet, finding another student with whom to work is a cinch—just look around your lab! Developing teamwork skills is crucial for later success in the workplace. Try applying teamwork to these tasks:

- Homework: Avoid blatant copying. Find other motivated students to work with, but beware of leeches, or people who seek only to benefit from your labors. Work alone, and then meet with your team to discuss results and stumbling blocks. Teaching one another will strengthen your own knowledge.

- Tests: Always study alone before joining a study group. Then, bring your questions to the group, and address stumbling blocks. Quiz each other. Above all else, stay focused. Give yourselves hourly breaks. And be careful! Group study can easily turn into mere socializing.

- Projects: With all projects, brainstorm, choose a leader, set firm schedules, and stay focused on goals—details should come later. Project work involves good group dynamics—how well people work together. Be aware that you will have difficult periods after going through the initial formalities. Although conflict yields better work, always respect one another's feelings. Remember always to share the burden and encourage each other.

7.1.4   Manipulating Graphics

To manipulate a graphical Maple object, such as a plot, select the image by pointing the mouse at it and clicking once with the left mouse button. A box surrounding the image will appear. You can now edit the selected object by performing the following tasks:

- *Preferences:* Select <u>F</u>ile→Preferences... and then click Plotting. I suggest you use in-line plots so that they appear in the worksheet.
- *Resizing:* Select and drag the black squares—called *handles*—that surround the highlighted plot.
- *Editing:* Select <u>E</u>dit or press the right mouse button to cut, paste, and copy.
- *Customizing plots:* Consult **?plot[options]** and **?plot3d[options]** for information on the options that most plotting functions accept. You may also use an interactive builder, menus, icons, and right clicks, as discussed in How To...Plot...Overview. See also **?plots[interactive]**, **?style2**, and **?style3**.

7.1.5   Saving and Printing Plots

By default, Maple includes plots inside worksheets. To save an in-line plot to a separate file outside of your MWS file, right click the plot and select <u>E</u>xport As, followed by your preferred file format. If you change your plotting settings in <u>F</u>ile→Preferences... to make plots appear in separate windows, you can use <u>F</u>ile menu options on the plots for saving and printing. You can also redirect plot output to other files from the command line. For instance, I have generated many plots in this text with the following commands:

```
> with(plots):
```

```
> plotsetup(ps,plotoutput='test.ps',plotoptions='height=3in,
        width=4in,color=grey,noborder,portrait'):
```

If you use my system, enter **plotsetup(inline)** to reset in-line plots. For more information on customizing output, consult **?plotsetup** and **?plot[device]**.

PRACTICE!

1. Plot the function x along $0 \le x \le 10$.
2. Change the axes to a box.
3. Resize the plot to about half the original size.
4. Cut the plot and insert it inside a new execution group.

7.2   TWO-DIMENSIONAL PLOTS

This section reviews further aspects of two-dimensional, or **2D**, plotting with **plot**.

7.2.1   Syntax

The **plot** function has the general syntax **plot(f,h,v,o)**, where

- **f** = an expression, such as a name or function
- **h** = a horizontal range expressed as **name=low..high**
- **v** = a vertical range expressed as **name=low..high**
- **o** = options expressed as the sequence **opt1=value,opt2=value,...**

For instance, to display a graph of $\sin x$ along $0 \le x \le \pi$, with the title Sine Plot printed on top, enter **plot(sin(x),x=0..Pi,title="Sine Plot")**.

7.2.2 Functions

The following statements produce the same plot of $y = f(x) = \sin x + 1$ using different forms of syntax:

Step 158: Examples of Plotting Functions

```
> plot(sin(x)+1,x=0..Pi);                     Directly plot an expression.
> y:=sin(x)+1: plot(y,x=0..Pi);          Plot an expression assigned to a name.
> f:=x->sin(x)+1: plot(f(x),x=0..Pi);           Directly plot a function.
```

Three equivalent plots are generated. To conserve space, only one is shown.

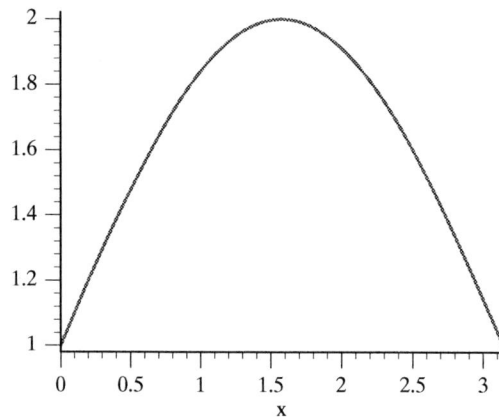

Never specify an entire equation in **plot**! Except for **restart**, the example in Step 159 is incorrect.

Step 159: Never Do This!

```
> plot(y=sin(x)+1,x=-10..10);                 Attempt to plot an equation.
Plotting error, empty plot                 plot does not accept equations!
```

If you wish to plot an equation, you can use an *implicit plot*, as discussed in Section 7.5.5. Consult **?plot[function]** for further rules, syntax, and shortcuts.

7.2.3 Ranges

Maple plots values of **f** for a given **h**:

- The **horizontal axis**, or *domain*, corresponds to the independent variable bounded by **h**.
- The **vertical axis**, or *range*, corresponds to the dependent variable evaluated as values of **f**. Supply a vertical-axis range **v** when plots seem too compressed or chopped.

For instance, to plot only the positive values of $\sin(x)$, you may enter either of the following inputs:

- **plot(sin(x),x=0..4*Pi,y=0..1)**
- **plot(sin(x),x=0..4*Pi,0..1)**

You can also specify infinite ranges with the symbolic constants **-infinity** and **infinity**. Consult **?plot[range]** and **?plot[infinity]** for more information.

7.2.4   Options

At first the amount of options might seem bewildering. I suggest that you initially use the interactive plot builder, as described in Graphics... Interactive Plot Builder and **?plots[interactive]**. Follow these steps for a simple example:

- Enter the expression **x** in an execution group, and press Enter.
- Select the output, which should just be x.
- Right-click the output and select the options Plots→Plot Builder.
- Ensure that the option 2-D plot is selected and click Next.
- Select the variety of options shown in Figure 7.1.
- Click Plot when you are finished.

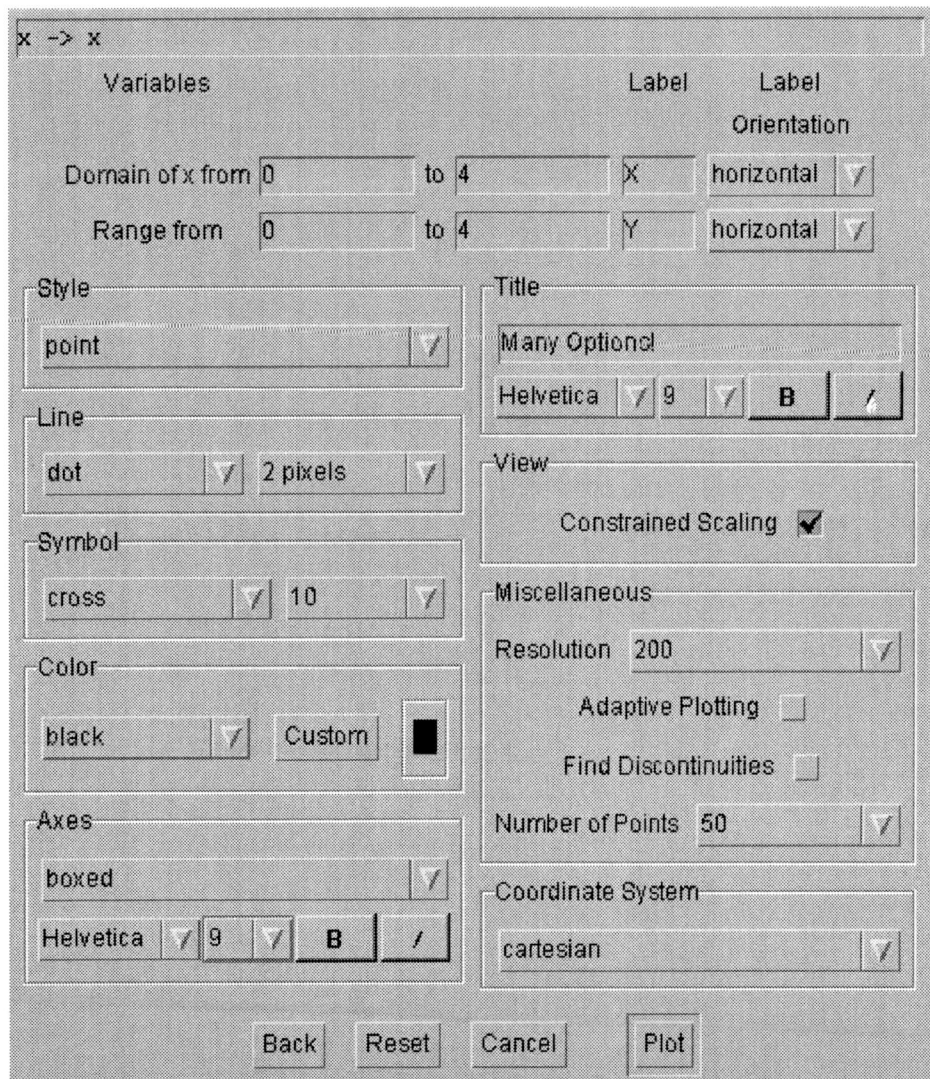

Figure 7.1   Interactive Plot Builder Options

Maple will actually convert all of the options you selected into Maple input and then enter the command to produce the plot just below the input. Step 160 demonstrates what you should see.

Step 160: Command-Line Plot Options

```
> plot(x, x = 0 .. 4, view = [DEFAULT, 0 .. 4], labels =
["X", "Y"], style = point, linestyle = DOT, thickness = 2,
color = black, axes = boxed, axesfont = [HELVETICA, 9],
title = "Many Options!", titlefont = [HELVETICA, 9], scaling
= constrained, adaptive = false);
```

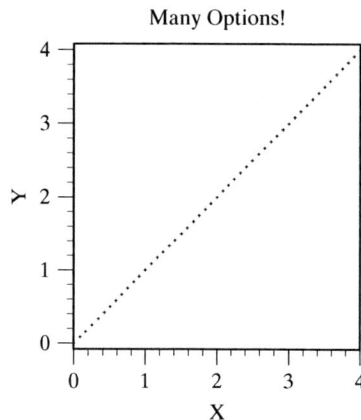

Many Options!

Also, try these context-menu selections (right-click):
Style→Point
Style→Symbol→Cross
Axes→Boxed
Projection→Constrained

To directly access the interactive plot builder, see **?plots[interactive]**. Eventually, you will learn the various plot options that are described in **?plot[options]**. To avoid tedious customizing, you may assign plot options **?setoptions**.

PRACTICE!

5. Plot the function $x = y^2 - 1$ for $0 \leq y \leq 1$. Which variable is independent? Dependent?

6. Plot the expression $\frac{1}{x}$ for $-1 \leq x \leq 1$. Hint: consider setting the **discont** option, since this curve has a discontinuity at 0.

7. Assign $f(x) = x \sin(x)$ using functional notation. Plot $f(x)$ for $0 \leq x \leq \pi$.

8. Plot the function $f(x) = e^{-x^2}$ for $-\infty \leq x \leq \infty$.

9. Create this plot:

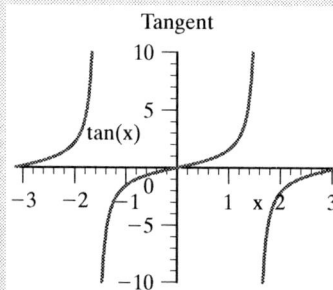

Tangent

7.3   PLOTTING MULTIPLE EXPRESSIONS

You can discover qualitative differences in models by plotting functions together. This section demonstrates different techniques for plotting multiple functions on the same graph.

7.3.1   Multiple Plots

Multiple plots contain graphs of different expressions in the general form of $y = f_1(x), y = f_2(x), \ldots$, as shown in Figure 7.2. In this example, each function $f_i(x)$ has the same independent variable x and is graphed with the same vertical axis y.

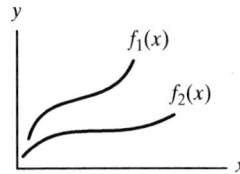

Figure 7.2   Multiple Plots

Enter `plot({f1, f2,...},h, v, opts)` for multiple function plots. For instance, suppose you wish to plot both $y = \sin x$ and $y = x - \frac{x^3}{3!} + \frac{x^5}{5!}$ on the same graph. Recall that $n!$ denotes a factorial: See **?!**. Try the following steps to create a multiple plot of these two functions:

Step 161: Multiple Plots

```
> x := 'x':                                                    Clear x.
```

```
> y1 := sin(x):                           Assign the function f₁(x) = sinx to y₁.
```
```
> y2 := x-x^3/3!+x^5/5!:           Assign the function f₂(x) = x - x³/3! + x⁵/5! to y².
```

Plot both $y=f_1(x)$ and $y=f_2(x)$ for $0\le x\le 2\pi$. Constrain the vertical range with the option $y = -1..1$:

```
> plot({y1,y2}, x=0..2*Pi, y=-1..1, tickmarks=[3,3]);
```

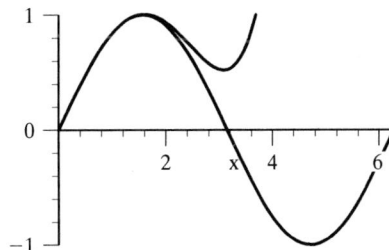

Both expressions are functions of x.

You should specify `[f1,f2,...]` and `opt = [opt1,opt2,...]` to apply different options to respective expressions. Also, consult **?plot[multiple]** for more details.

7.3.2 Superimposing Plots

Maple stores plots as ***plot structures***, which are data that determine all plot characteristics. You can output a plot structure by assigning a name when calling **plot**. Using the assignments *y1* and *y2* in Step 161, try the following steps:

Step 162: Plot Structures

```
> P1 := plot(y1,x=0..2*Pi,y=-1..1);
```
A REALLY BIG Maple plot structure appears and is omitted to conserve space.

Assign a plot structure to P1. Next time, terminate the input with a colon to suppress the output!

Next, you will need to use the **display** function, which is stored in the **plots** library package. To access **display**, you will need to make the **plots** library accessible:

Step 163: Accessing **plots** Library Package

```
> with(plots);
```
You will see many functions also omitted to conserve space.

*Make the functions in the **plots** library package accessible.*

Next time, use a colon to suppress the output!

You can now use every function contained in **plots** until you restart or end your session. To display the plot, enter **display(P1, *opts*)**:

Step 164: Display Single Plot

```
> display(P1);
```
Maple displays a sine plot that is omitted to conserve space.

*Evaluate the plot structure P1. Also, try entering just **P1**.*

Now, redo Step 161 by using plot structures and the **display** function:

Step 165: Plot Structure

```
> P1 := plot(sin(x),x=0..2*Pi,y=-1..1):
> P2 := plot(y2,x=0..1,y=-1..1):
```
Assign a plot structure to the name P1.

Assign a plot structure to the name P2.

Display multiple plots with **display({*P1, P2,...*}, *opts*)**, where **{P1, P2, ...}** represents a set of plot structures:

Step 166: Display Multiple Plots

```
> display({P1,P2},tickmarks=[3,3]);
```
Display both functions in one plot.

Maple displays a plot that resembles the multiple plot in Step 161 and is omitted to conserve space.

For more information, consult **?plot[structure]** and **?plot[display]**.

PRACTICE!

> 10. Plot both $f_1(x) = x + 1$ and $f_2(x) = -x - 1$ for $-2 \leq x \leq 2$ on the same graph.
> 11. Assign plots for $y = x + 1$ and $y = -x - 1$ to A and B, respectively, along $-2 \leq x \leq 2$.
> 12. Superimpose and display both plots stored in A and B with **display**.

7.3.3 Parametric Plots

Consider the pair of functions $x(t) = \cos(t)$ and $y(t) = \sin(t)$. Both $x(t)$ and $y(t)$ are called **parametric functions** because each function is a function of the same parameter. Consequently, each parametric function determines the coordinate value of a different axis. How? In the example in this section, different values of the parameter t produce pairs of x and y coordinates on an xy-plot. For instance, when $t = 0$, $(x, y) = (1, 0)$. When $t = 0.1$, $(x, y) = (0.995, 0.0998)$, and so forth.

To plot horizontal and vertical parametric functions, **fh** and **fv**, for parameter **t**, enter **plot([fh, fv, t=range], opts)**:

Step 167: Parametric Plots

```
> x:=cos(t): y:=sin(t):
```
Assign two parametric functions.

```
> plot([x,y,t=0..Pi/2]);
```
$\cos(t)$ *is the horizontal coordinate.* $\sin(t)$ *is the vertical coordinate.*

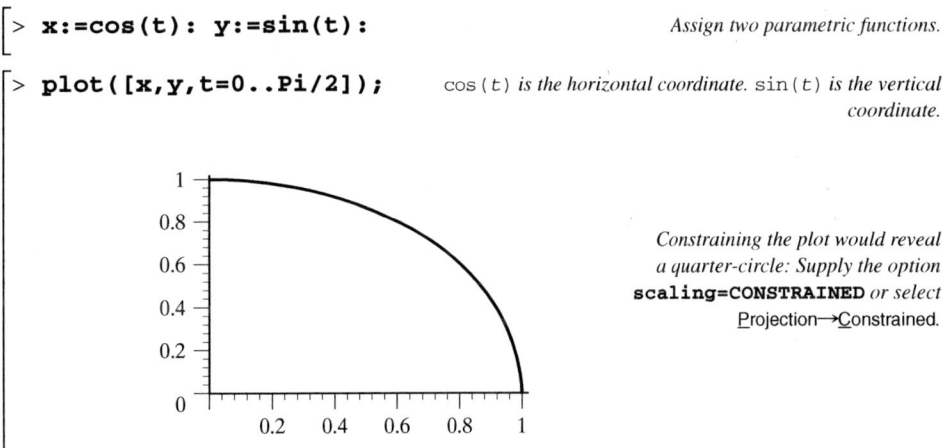

Constraining the plot would reveal a quarter-circle: Supply the option **scaling=CONSTRAINED** *or select* Projection→Constrained.

Never confuse multiple function plots with parametric function plots:

- multiple functions: **plot([funcs],h)**, where **h** is a horizontal range.
- parametric functions: **plot([funcs,t])**, where **t** is a parametric variable.

Do you see how the two syntaxes differ? Consult **?plot[parametric]** for more options.

PRACTICE!

> 13. Plot $x = \cos(t)$ and $y = \sin(t)$ for $0 \leq t \leq 2\pi$. What shape does this graph have?
> 14. Plot both $f(t) = \cos(t)$ and $g(t) = \sin(t)$ for $0 \leq t \leq 2\pi$ on the same graph. What is the difference between this graph and that of the previous problem?

7.4   THREE-DIMENSIONAL PLOTS

This section introduces aspects of three-dimensional, or **3D**, plotting.

7.4.1   Syntax

Many two-dimensional plot commands have 3D versions. Try appending **3d** to a 2D function name to find its 3D version. For instance, for functions expressed in the form $z = f(x, y)$, enter **plot3d(*expr,x=range,y=range,opts*)** to plot values of **z**. In general, **plot3d** plots a surface above and below the xy-plane:

Step 168: Plotting 3D essions

```
> opts := axes=FRAME,shading=zgrayscale,title="3D PLOT":
```

```
> plot3d( cos(x)^2+sin(y)^2,x=0..2*Pi,y=0..2*Pi, opts );
```

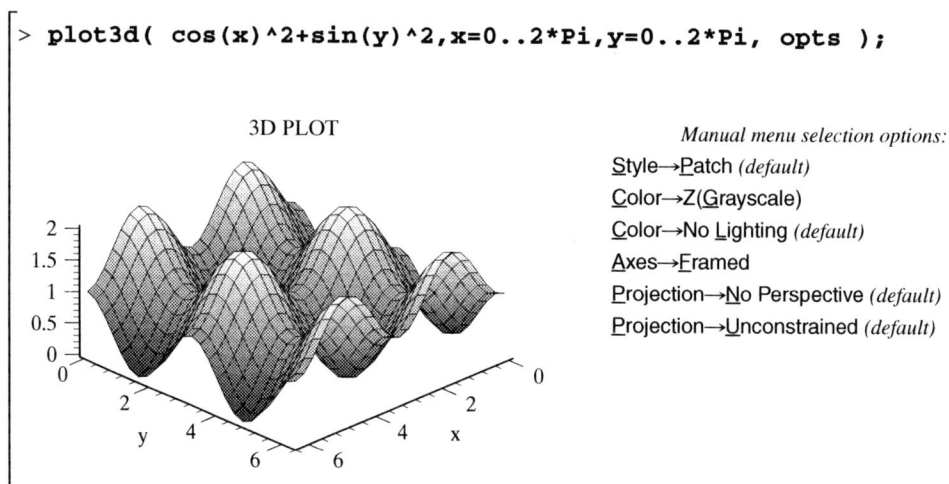

3D PLOT

Manual menu selection options:
Style→Patch *(default)*
Color→Z(Grayscale)
Color→No Lighting *(default)*
Axes→Framed
Projection→No Perspective *(default)*
Projection→Unconstrained *(default)*

7.4.2   Customizing

See Basic Features...Plots...modify 3-d plot... for help on various techniques. As with 2D plots, you use the interactive plot builder: try entering **interactive (cos(x)^2+sin(y)^2)**. You may also right-click the plot to pull up context menus that help you set these options, as shown in Step 168. Because I cannot show color here, I chose the **shading=zgrayscale** option, but you have many more choices. Instead, in Step 168, set **shading=zhue** to get a more interesting picture. To find more options available for 3D plots, see **?plot3d[options]** and **?style3**. See also **?plots [interactive]**, **?setoptions3d**.

7.4.3   Orientation

Orientation is an important option of 3D plots that measures how plots rotate about each axis. (See Figure 7.3.) After you select a plot, the following symbols appear on the Tool Bar:

- ϑ (cursive *theta*) = rotation in the xy-plane
- φ (cursive *phi*) = rotation from the z-axis

Both ϑ and φ represent angles that you can modify. Therefore, to change a 3D plot's orientation, perform the following:

- Select the plot.

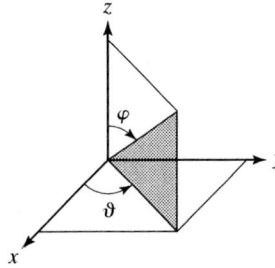

Figure 7.3   3D Orientation

- Hold the left mouse button and move the mouse. The plot will slowly spin.
- Monitor the angles next to ϑ and φ on the Tool Bar.

You may retain these angles by entering the option **orientation=[a,b]**, where $\vartheta = $ **a** and $\varphi = $ **b**:

Step 169: Changing Orientation

```
> with(plots):
```
If you previously entered **restart***, you lost your access to* **plots***.*

```
> opts := axes=BOXED, tickmarks=[2,2,2],
          labels=["x","y","z"], grid=[5,5]:
```
Choose options.

```
> P := plot3d(exp(x*y),x=0..1,y=0..1,opts):
```
Assign a 3D plot.

```
> display3d(P); display3d(P,orientation=[145,75]);
```
Display plots.

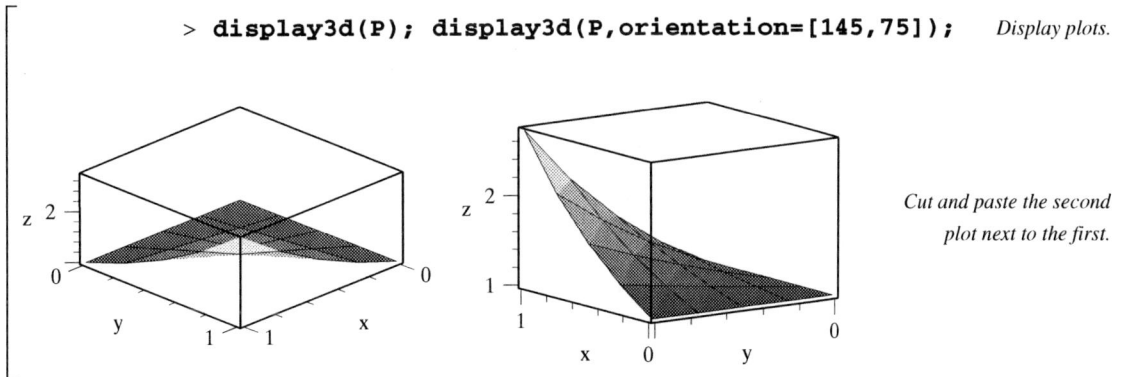

Cut and paste the second plot next to the first.

PRACTICE!

15. Plot $f(x, y) = e^{(x^2 - y^2)}$ for $-1 \le x \le 1$ and $-1 \le y \le 1$.
16. Describe the surface of $f(x, y) = x + y$.
17. Plot $f(x, y) = \ln(x - y^2)$, for $1 \le x \le 1.5$ and $1 \le y \le 1.1$. Set appropriate options to see "drip-marks."

7.5 MISCELLANEOUS

This section reviews other common Maple plotting functions.

7.5.1 Legends

When you have multiple plots, you might want to help distinguish the plots by providing an explanation of the different lines, which is called a ***legend***. To generate a plot legend, use the plot option **legend=[*string1*, *string2*,...]** when plotting **[*expr1*, *expr2*,...]**, where you provide a string for each expression:

Step 170: Plot Legends

```
> opts := legend=["straight","curved"], linestyle=[1,3],
        labels=["",""], tickmarks=[0,0], style=[LINE,POINT],
        thickness=3, axes=boxed:
```

```
> plot([x,x^2],x=0..5,opts);
```
                                                            Plot multiple expressions.

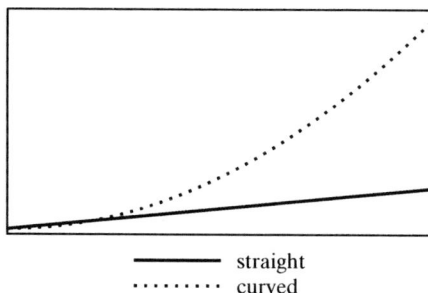

    ——— straight
    .... curved

To modify the legend, you may either change your options or use a context menu, as discussed in **?legend**.

7.5.2 Text

Besides titles and legends, you can annotate plots by plotting text strings. To plot string at horizontal and vertical coordinates **x** and **y**, respectively, enter **textplot([*x*, *y*, *string*])**:

Step 171: Text Plot

*Enter **with(plots)** if you have not already done so because textplot belongs to the plots library package.*

```
> T:=textplot([Pi/2,3/2, "maximum"]):
```
                                        Plot the string "maximum" at $x = \pi/2$
                                                        and $y = 3/2$.

Step 172: Combine Text and Graphical Plots

```
> opts:=tickmarks=[3,2]:
```
Assign plotting options.

```
> G:=plot(sin(x),x=0..Pi,opts):
```
Assign a graphical plot.

```
> display(G,T);
```
Superimpose the sine graph with the text plot.

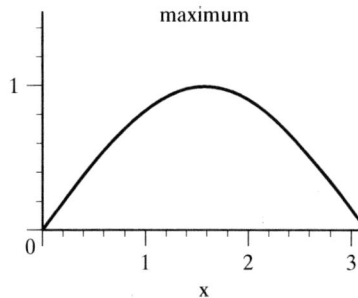

To vertically and horizontally align **text** about **x** and **y**, use the option, **align={vert, horz}**:

- Enter **ABOVE** or **BELOW** for **vert**.
- Enter **RIGHT**, **CENTER**, or **LEFT** for **horz**.

You may change fonts with the **font=[FONT, STYLE, SIZE]** option. Consult **?plots[textplot]** and **?plots[textplot3d]** for more information.

7.5.3 Points

In Maple, a point is represented in two dimensions with the notation $[x, y]$. For instance, you would enter the coordinate $(x, y) = (1, 2)$ as **[1,2]**. To plot a collection of points as a line, collect n data points inside a list of lists $[[x_1, y_1], [x_2, y_2], \ldots, [x_n, y_n]]$ and use the **plot** command:

Step 173: Plotting Points

```
> L := [[1,2],[2,1]];
```
Assign a list of points. Each point is also a list $[x_i, y_i]$.

```
> plot(L,thickness=3);
```
Plot the points. Maple automatically creates lines between the points.

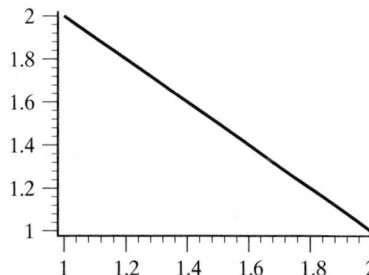

For more information, consult **?plots[pointplot]**, **?plottools[point]**, **?geometry[point]**, and **?geom3d[point]**.

7.5.4   Coordinate Systems

Cartesian graphs plot functions along mutually perpendicular x-, y-, and z-axes. However, other coordinate systems provide better visual tools for certain functions. For instance, *cylindrical coordinates* help model circular objects:

Step 174: Coordinate Systems

```
> opts := coords=cylindrical,style=wireframe:
```
Choose cylindrical coordinates.

```
> plot3d(1,theta=0..2*Pi,z=0..1,opts);
```
Specify $r = 1$ for $0 \leq \varphi \leq 2\pi$ and $0 \leq z \leq 1$.

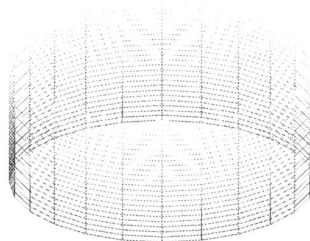

Consult **?coords**, **?plot[coords]**, and **?changecoords** for further description and information on many related functions.

7.5.5   Implicit Plots

Maple provides functions for plotting entire equations. For instance, let $F(x, y)$ represent the entire equation $y^3 + xy = 1$. $F(x, y)$ is called an ***implicit function***, which is a function of mutually dependent variables. Implicit functions simultaneously vary x and y values to satisfy the equation, now represented by F. To implicitly graph ***F***, enter **implicitplot(*F, h, v, opts*)**:

Step 175: Implicit Plots

*Enter **with(plots)** if you have not already done so because **implicitplot** belongs to the **plots** library package.*

```
> implicitplot(y^3+x*y=1,x=-10..10,y=-3..3);
```
Plot $F(x, y) = y^3 + xy = 1$.

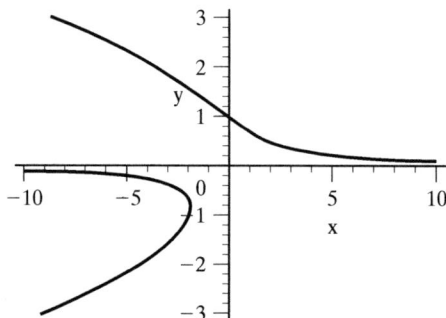

Consult **?plots[impliciplot3d]** for 3D implicit plots. To bypass implicit plots, you may also wish to investigate **?smartplot** and **?plot_real_curve**.

7.5.6   Geometric Objects

Maple provides library packages for drawing geometric objects:

- **plots**: Graph polygons with **polygonplot** and **polygonplot3d**.
- **plottools**: Construct many common geometric shapes and modify other plots.
- **geometry**: Define and assign 2D shapes. Consult **?geometry[draw]** for graphics.
- **geom3d**: Define and assign 3D shapes. Consult **?geom3d[draw3d]** for graphics.

For instance, suppose you were curious what a *Great Stellated Dodecahedron* looks like. To find out, you would enter the following:

Step 176: Drawing Geometric Objects

Enter **with(plots)** *if you have not already done so because* **polyhedraplot** *belongs to the* **plots** *library package.*

```
> polyhedraplot([2,2,2], polytype=GreatStellatedDodecahedron,
               style=PATCH, scaling=CONSTRAINED);
```

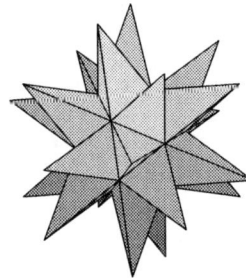

Pretty cool, huh? This plot looks even better in color!

7.5.7   Animated Plots

Maple can display functions that vary with time t as an ***animation***, which is a sequence of images that simulate motion. Animated plots flash a sequence of changing "snapshots" that simulate actual motion for different values of time t. Time-dependent functions, such as $f(x, t) = x \sin(xt)$, produce functions that change in space x as well as time t. Note that time-dependent functions, like $f(x, t)$, have three overall coordinates:

- Two spatial coordinates, which typically are represented as x and y. Spatial coordinates are plotted on horizontal and vertical axes.
- One temporal coordinate, which typically is represented as time t. As time changes, Maple redraws (x, y) coordinates for new values of t.

For instance, enter **animate(f,x=x1..x2,t=t1..t2)** to animate $f(x, t)$ where $x_1 \le x \le x_2$ and $t_1 \le t \le t_2$:

Step 177: Animation

Enter **with(plots)** *if you have not already done so because animate is stored inside the plots library package.*

```
> animate(x*sin(x*t),x=0..10,t=0..2*Pi);
```
Animate the function
$f(x,t) = x\sin(xt)$
for $0 \le x \le 10$ *and* $0 \le t \le \pi.$

A plot appears surrounded by boxes.

Select the plot. Then, select Animation→
Play. *Also, try the PLAY icon.*

The Tool Bar icons resemble music-player icons. You may select an animated plot and then press the play icon ▶, stop icon ■, and other icons. To take instantaneous "snap-shots" of a plot at various times without animating the entire plot, supply display with the **insequence=false** option:

Step 178: Animated Snap Shots

Enter **with(plots)** *if you have not already done so because* **animate** *and* **display** *belong to* **plots** *library package.*

```
> opts:=axes=NONE,labels=[" "," "],frames=6:
```
Take six "snapshots."

```
> P:=animate(x*sin(x*t),x=0..10,t=0..100,opts):
```
Store the animated plot.

```
> display(P,insequence=false);
```
insequence=false *tells Maple not to animate in real-time.*

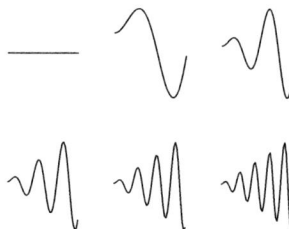

For more details, consult **?plots[animate]**, **?plots[animate3d]**, **?contextanimate**, and **?plotinterface [animate]**.

PRACTICE!

18. Plot the functions $f(x) = x$ and $f(x) = x^2$ for $0 \le x \le 10$ on the same graph. Label each function with a legend and then text plot.

19. Plot a triangle. Use the coordinates $(0, 0)$, $(1, 1)$, and $(2, 0)$.

20. Plot $r(\theta, z) = z - \theta^2$ for $0 \le \theta \le 2\pi$ and $-1 \le z \le 1$. Hint: Consult **?plot3d[coords]**.

21. Plot a circle $x^2 + y^2 = 1$ using an implicit plot.

22. Plot a circle with origin $(1, 1)$ using Maple's **plottools** package. Hint: Consult **?circle**.

23. Animate $f(x, t) = \tan(xt)$ for $0 \le x \le 100$ and $0 \le t \le 1$. Display only 10 frames. Restrict the vertical height to $-20 \le y \le 20$.

7.6   APPLICATION: PIPELINE FLOW

This section demonstrates Maple's plotting capabilities using a pipeline flow example.

North America has vast reserves of natural gas, the cleanest-burning fossil fuel. Natural gas is used by people, organizations, and industries for many purposes, such as heating and generating power. Over 1.3 million miles of underground pipes deliver natural gas throughout the United States. ANR Pipeline Company is one the nation's largest providers of natural gas transportation and storage service. ANR operates approximately 10,600 miles out of the 200,000 miles of large-diameter interstate pipeline. Courtesy of ANR Pipeline Company, subsidiary of The Coastal Corporation.

7.6.1   Background

Pipelines carry water, gas, oil, and effluents. Not only do the characteristics of the flowing substance affect flow, but other factors, such as flow rates, pipe roughness, elevation changes, and bends, can greatly change flow behavior as well. While theoretical approaches to the analysis of flow behavior exist, many empirical techniques also greatly assist such analysis. For instance, the Weymouth equation predicts reasonable values for the flow rate Q of natural gas in a pipe, where C_Q = a constant, T_b = temperature base, P_b = pressure base, D = internal diameter of the pipe, G = relative gas density, T_f = average flowing temperature, L = length, P_1 = upstream pressure, and P_2 = downstream pressure:

$$Q = C_Q \frac{T_b}{P_b} D^{\frac{8}{3}} \left(\frac{P_1^2 - P_2^2}{GT_f L} \right)^{\frac{1}{2}} \tag{7.1}$$

7.6.2 Problem

Given $C_Q = 0.0037477$, $T_b = 293$ K, $P_b = 1.060$ Kg/cm^2, $D = 305$ mm, $G = 0.600$, $T_f = 300$ K, and $L = 130$ km, describe the relationship between Q, P_1, and P_2.

7.6.3 Methodology

Although Eq. (7.1) seems complicated, using a plot should clarify the behavior of the "main" parameters. So, first restart Maple:

Step 179: Pipeline Flow—Initialize

> **restart:**   *Restart your Maple session.*

For the application, first assign the model to keep the equation free of variable assignments for later study:

Step 180: Pipeline Flow—Model

> **Q:=Cq*(Tb/Pb)*D1^(8/3)*((P1^2-P2^2)/(G*Tf*L))^(0.5);**

$$Q := \frac{CqTbD1^{\left(\frac{8}{3}\right)}\left(\dfrac{P1^2 - P2^2}{GTfL}\right)^{.5}}{Pb}$$

Assume that flow rate is a function pressure.

Now, assign pertinent data:

Step 181: Pipeline Flow—Separate and State

> **Cq:=0.0037477:Tb:=293:Pb:=1.06:D1:=305:G:=0.6:**
> **Tf:=300:L:=130:**

Check the expression for Q:

Step 182: Pipeline Flow—Check

> **evalf(Q,2);**   *Calculate with two floats and evaluate Q.*

$$Q := 93000.305^{\left(\frac{2}{3}\right)}(.000044P1^2 - .000044P2^2)^{.5}$$

Now, you can plot the function. Because Q is a function of two independent variables, use **plot3d** to create a three-dimensional plot. Whereas the pressures are represented by the "bottom" axes, the height of the plot represents Q. You might need to adjust the ranges for $P1$ and $P2$ to display relevant features that illustrate the model's physical aspects:

Step 183: Pipeline Flow—Solve and Report

> **plot3d(Q, P1=100..120, P2=0..100, title="Pipe Flow",**
>   **axes=BOXED, gridstlye=RECTANGULAR, orientation=[135,**
>   **65], tickmarks=[2,2,4]);**

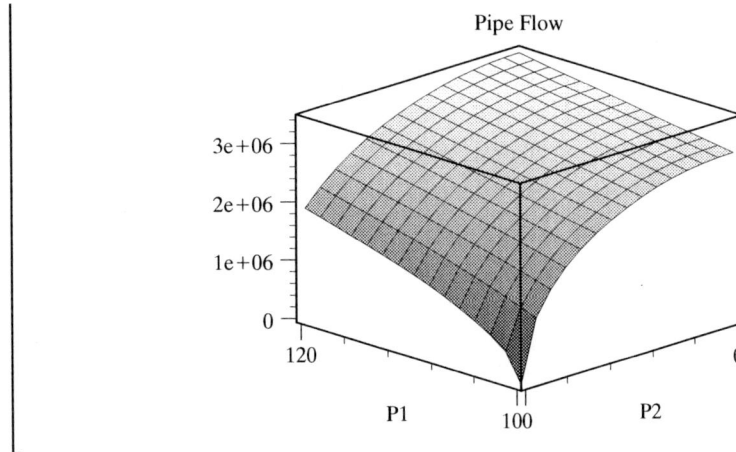

Pipe Flow

7.6.4 Solution

The plot shows that large pressure differences generate large flow rates. Differences in pressures, called *pressure drops*, drive pipe flow, as shown in the plot. Therefore, as the pressure difference decreases, the flow rate diminishes, all other factors equivalent.

KEY TERMS

2D	3D	Cartesian graphs
horizontal axis	implicit function	legend
multiple plots	orientation	parametric functions
plot structures	plot	vertical axis

SUMMARY

- Maple offers a variety of graphics tools, as described in Graphics....
- You can find help on most tools you need in **?plot** and **?plot3d**.
- Two-dimensional plots (**plot**) graph expressions on horizontal and vertical axes.
- You can supply a variety of options to **plot**, as described in **?plot[options]**.
- You may assign a plot structure to a name.
- A set or list of expressions will plot as a multiple plot.
- The **display** function will plot multiple plot structures.
- You modify a 3D plot's orientation along all three axes.
- The **plots** library package contains many useful plotting functions.

Problems

1. What is a Maple plot?
2. Demonstrate how to apply a vertical range to a two-dimensional plot.
3. What is a coordinate system?
4. Why do you think that Maple does not require a label if you specify a vertical range with plotting?

5. Give an example of all options that are listed in **?plot[options]**. For example, **plot(x+1,x=0..1,title="I am an example!")** demonstrates the title option.

6. Plot the functions $y = \cos t$ and $y = \sin^2 t$ on the same graph for $0 \le t \le \pi$.

7. Plot the parametric functions $x = \cos^2 t$ and $y = \sin^2 t$ for $0 \le t \le \pi$.

8. How do multiple plots and parametric plots differ?

9. Plot a circle of radius 1 and origin $(0, 0)$:

    a. Use **implicitplot** for the equation $x^2 + y^2 = 1$.

    b. Use **smartplot** for the equation $x^2 + y^2 = 1$.

    c. Use a parametric plot.

    d. Use **circle**. (Hint: Remember to first load **plottools**.)

10. The Gaussian-distribution, or normal-distribution, function is popular in statistical models. For example, we may have

$$f(x) = \frac{1}{\sigma\sqrt{2\pi}} \exp\left(\frac{-(x - m)^2}{2\sigma^2}\right), \tag{7.2}$$

where $\sigma > 0$ and $-\infty < x < \infty$. Plot $f(x)$ for standard deviation $\sigma = 7.5$ and mean $m = 65$. Hint: Pick reasonable ranges to produce a clear plot.

11. Repeat the problem in Section 7.6, but use only 50 Kg/cm^2 for P_2. Plot the function $100 \le D \le 400$ mm and the given variation of P_1.

12. Plot the functions $f(t) = \sin(t)$ and $f(t) = \sin(t + \theta)$ on the same plot for $\theta = 45°$. Let $0 \le t \le 10$ and label both plots. What effect does θ have on the second plot?

13. Plot the results of Problem 9 in Chapter 6 for values of $h, b_1,$ and b_2 that yield a trapezoidal area of 10. Hints: Use an implicit plot with **implicitplot3d** because the function has three independently ranging variables. Remember to load the **plots** library package! Choose ranges such as $0 \le b_1 \le 20$, $0 \le b_2 \le 20$, and $0 \le h \le 20$, though you might wish to try larger spans that demonstrate the function's features a bit better.

14. Due to the centripetal force on vehicles, curved roadways are banked at a *superelevation angle* ϕ. You can calculate the necessary superelevation for a bank using the formula

$$\tan(\phi) = \frac{v^2}{gr}, \tag{7.3}$$

where v is the vehicle's velocity and r is the radius of curvature of the roadway. Given $g = 32.2$ ft/s^2, $45 \le v \le 75$ miles per hour, and $500 \le r \le 1000$ ft, plot ϕ. Be sure to rotate the plot to illustrate important features.

15. Given elevation above sea level E (m) and air temperature t_a (°C), you can compute air density ρ_a (kg/m^3) as

$$\rho_a = \frac{353(1 - 0.2257 \times 10^{-4}E)^{5.255}}{t_a + 273}. \tag{7.4}$$

Plot ρ_a with respect to both E and t_a. Hint: Use **plot3d**. Explain how changes to both elevation and air temperature qualitatively affect air density.

16.   Suppose you are designing a rectangular room with only enough supplies to handle 100 m^3. Given a fixed ceiling height of 2 m, what dimensions of the room provide the maximum perimeter? Use graphical techniques to determine your answer.

17.   Draw a "smiley face," as shown in Figure P7.4, using Maple.

Figure P7.4   Smiley Face

Hints: Store each portion as a separate plot structure. Enter `display({P1, P2,P3,P4},options)` to display the entire plot. Do not forget to load the `plots` library package beforehand! Use circles for the face and eyes. An arc or a pair of parametric functions can define the smile. Consult `?circle` and `?arc`. Be sure to enter `with(plottools)` before using `circle` or `arc`! For plot options, set `axes=NONE`.

18.   Consider the plot of experimentally measured data shown in Figures 5.2 and 5.3. How would you determine the most appropriate line to draw through the scattered points? A common method for determining such lines is the least-squares formula for n pairs of dependent and independent variables, x and y, namely,

$$y = a_1 x + a_0, \tag{7.5}$$

where

$$a_1 = \frac{(n)\left(\sum_{i=1}^{n} x_i y_i\right) - \left(\sum_{i=1}^{n} x_i\right)\left(\sum_{i=1}^{n} y_i\right)}{(n)\left(\sum_{i=1}^{n} x_i^2\right) - \left(\sum_{i=1}^{n} x_i\right)^2} \tag{7.6}$$

and

$$a_0 = \frac{\left(\sum_{i=1}^{n} y_i\right)}{n} - (a_1)\left(\frac{\sum_{i=1}^{n} x_i}{n}\right). \tag{7.7}$$

Using these equations, solve the following problems:

(a) Assign the data in Table 7.1 for u and w using sequences. For instance, you could assign `DP:=1,2,3,4,5`, though unnecessary.

TABLE 7.1   Mass-Spring Experimental Data

Data Point	1	2	3	4	5
w (Newtons)	0.00	5.00	10.0	20.0	50.0
u (cm)	0.00	0.51	0.98	1.21	5.16

(b) Investigate Maple's **add** function. Why can this function help perform the sums inside the least-squares formula? Hint: try **add(DP[i], i=1..5)**.

(c) Evaluate a_1 and a_0.

(d) Assign the least-squares formula, using w, u, a_1, and a_0 as variables. Is your result linear or nonlinear?

(e) Assign a plot of the individual data points to a plot structure.

(f) Assign a plot of your least-squares formula to another plot structure.

(g) Display both plot structures superimposed together. Hint: Use **display**.

(h) How does the point plot relate to the line plot of the least-squares formula?

(i) Find Maple's statistics library package for performing a least-square fit. Hint: Look inside **?stats** for a linear-regression function.

(j) Evaluate the least-squares formula using Maple's statistics functions. Compare Maple's formula to the one that you generated in Problem 18d.

8

Substituting, Evaluating, and Solving

8.1 INTRODUCTION

An equation has the form **expr1=expr2**, but you may have simulated many equations with assignments by using **name:=expr**. For instance, you could simulate the equation for a line $y = mx + 6$ by assigning the expression $mx + b$ to y:

OBJECTIVES

After reading this chapter, you should be able to

- Substitute values into expressions
- Evaluate expressions
- Solve equations for symbolic values
- Solve equations for numerical values
- Verify solutions

Step 184: Expression Assignments

```
> restart;
```
Clear all assignments.

```
> y := m*x+b:
```
Assign an expression.

```
> x:=1: m:=2: b:=0: y;
```
$$2$$
Assign parameter values.
Now, the expression is evaluated.

You may also use functional notation **f:=name->expr** to simulate an equation:

Step 185: Functional Notation

```
> restart;
```
Clear all assignments.

```
> f := (x,m,b)→m*x+b:
```
Assign an expression using functional notation.

```
> f(1,2,0);
```
$$2$$
Select parameter values.
Now, the expression is evaluated.

In both cases, varying of the parameter values produces new y values. Unfortunately, these methods become cumbersome for larger problems. To ease your input and manipulation of expressions, this chapter reviews **substitution**, **evaluation**, and **solving** techniques.

PROFESSIONAL SUCCESS: TECHNICAL PRESENTATIONS

Presentations are common for many professions. Does public speaking make you nervous? If so, you should consider some of the following suggestions to calm your nerves:

- Preparation: Treat presentations as you would written reports: *Brainstorm, outline, write*, and *rewrite*. Break larger tasks into smaller tasks if the work seems daunting.

- Organization: **Tell them what you'll tell them**. Commence with a title, abstract, and overview. Next, provide an introduction and some background material. **Tell them**. Focus the scope of the problem. Include relevant theory, methodology, experiments, and examples. Compare expected and actual results. **Tell them what you told them**. Summarize your presentation. Discuss conclusions and recommendations. Always leave enough time for questions.

- Material: Never write down everything that you will say. Instead, summarize important concepts and points with brief bulleted statements, and spruce up your talk with many graphics. Also, people should actually be able to see the writing on such displays! Avoid having to say, "Well, you probably can't see this, but it says, ..."

- Style: Look up, and look around. If you find it helpful, pretend that everyone is wearing nothing but underwear. Never memorize every word. Speak slowly and calmly. Avoid nervous habits such as scratching or nail biting, and remember, it's important to smile.

8.2 SUBSTITUTION

Expressions are composed of subexpressions. Substituting for these subexpressions with other expressions helps manipulate equations, as introduced in this section.

8.2.1 Syntactic Substitution

Syntactic substitution replaces operands of expressions with other expressions. To substitute each occurrence of **old** with **new** inside **expr**, enter **subs(old=new, expr)**. Both **old** and **new** are often individual names, as demonstrated next:

Step 186: Syntactic Substitution

> `restart:`   *Clear variable assignments.*

> `expr := (x*y^2)*sin(x*t);`   *Assign an expression.*

$$expr := xy^2 \sin(xt)$$ *Ensure that you first entered* **restart**.

> `subs(x=a,expr);`   *Replace x with the new expression a.*

$$ay^2 \sin(at)$$ *Maple swapped x for a.*

Recall that Maple considers *equations* as expressions that include the equals sign (**=**). Therefore, The input ***old=new*** forms an expression of type *equation*. Hence, I can state an equivalent syntax for **subs** as **subs(eqn,expr)**, which you might see sometimes. For more information, consult **?subs**, **?alias**, **?subsop**, and **?trigsubs**.

8.2.2 Assignments

What happens to expressions **x** in Step 186? If you check the value of **x**, you will see that it remains unchanged:

Step 187: **subs** Uses Local Variables

> `x;`   *Check the value of* **x**.

$$x$$ *Using* **subs** *in Step 186 did not alter* **x**!

Why does **subs** not alter the expression or substituted variables? It is because **subs** uses *local variables*, which are variables defined "inside" a particular function and not used by other functions. Therefore, local variables do not carry global assignments. Most variables that you enter at the prompt are called ***global variables***. A variable that you assign at a worksheet prompt is global, because all future expressions that you enter will "know" about that variable and its assigned expression. So, what happens to a local variable? When a function that uses a local variable finishes evaluating, Maple resets the local variable to an unassigned state. Thus, Maple treats **x** and **a** as local during the substitution. After the substitution, the worksheet will not know about the temporary change from **x** to **a**.

What about the entire expression **expr**? You will see that **expr** also does not change:

Step 188: **subs** Does Not Alter Expression Assignments

> `expr;`   *Check the value of* **expr**.

$$xy^2 \sin(xt)$$ *Using* **subs** *in Step 186 did not alter* **expr**!

When you enter **subs(eqn,expr)**, you have actually entered just another expression that consists of function (**subs**) and two expressions (**eqn** and **expr**). An expression statement simply activates full evaluation without changing any assignments! If you wish to retain your substitutions, you need to assign the result to a name, which may include the name you used for **subs(*eqn*, *expr*)**:

Step 189: Assigning Results of Substitution

```
> restart:                                    Clear variable assignments.
```

```
> expr := (x*y^2)*sin(x*t):                   Assign an expression.
```

```
> subs(x=a,expr): expr;     Replace x with the new expression a. See if expr changed.
```

$$xy^2\sin(xt)$$

No! Remember that you must actually assign the result to a name.

```
> expr := subs(x=a,expr);          Replace x with the new expression a and
                                   assign the result to expr.
```

$$expr := ay^2\sin(at)$$

Maple performed the substitution and stored the result in expr.

8.2.3   Evaluation and Simplification

In general, **subs** does not evaluate expressions:

Step 190: Substitution Does Not Evaluate!

```
> sin(Pi/3);                                  Evaluate sin π/3.
```

$$\frac{1}{2}\sqrt{3}$$

Maple evaluated the expression.

```
> subs(x=Pi/3,sin(x));              Substitute x = π/3 into sin x.
```

$$\sin\left(\frac{1}{3}\pi\right)$$

Maple did not evaluate the expression.

However, **subs** still automatically simplifies expressions:

Step 191: Substitution Performs Automatic Simplification

```
> subs(x=1,x+x);          Arithmetic expressions are automatically simplified.
```

$$2$$

Maple simplified 1 + 1.

8.2.4   Multiple Substitutions

To perform multiple substitutions in which you exchange more than one expression, you could enter **subs({*eqns*},{*exprs*})**. Both *eqns* and *exprs* can consist of a sequence or single input, as follows:

- **{eqns}**, or **{eqn1, eqn2, ...}**, is a set of simultaneous substitutions.
- **{exprs}**, or **{expr1, expr2, ...}**, is a set of expressions provided for the substitutions.

Note that each **eqn** inside **{eqns}** takes the form **old=new**, where the **new** expression replaces the **old** expression. For instance, you may simultaneously substitute $a = 1$ and $b = 2$ into the expressions $a + b$ and ab:

Step 192: Multiple Substitutions

```
> subs({a=1,b=2},{a+b,a*b});
```
Substitute $a = 1$ *and* $b = 2$ *into the expressions* $a + b$ *and* ab.

$$\{2, 3\}$$

Sets have no order, so you could have entered **[a+b,a*b]** *to produce* [3, 2] *instead of* {2, 3}.

8.2.5 Algebraic Substitution

Substitution sometimes fails because Maple attempts substitution for surface operands and individual names. To see an instance of this failure, try substituting for a more complicated subexpression:

Step 193: Substitution Sometimes Fails

```
> expr:=(x*y^2)*sin(x*t):
```
Assign an expression.

```
> subs(x*y=1,expr);
```
Substitute $xy = 1$ *into* $xy^2 \sin(xt)$.

$$xy^2 \sin(xt)$$

subs *failed!*

Why did **subs** fail? Investigate the failure by displaying the main operands of $xy^2 \sin(xt)$ with **op**:

Step 194: Checking Operands

```
> op(A);
```
Display operands of an expression's surface type.

$$x, \ y^2, \ \sin(xt)$$

xy *does not explicitly appear inside* **expr** *as an operand.*

The expression xy does not explicitly appear in the expression, and thus, **subs** fails. Try *algebraic substitution* with the **algsubs** function. Algebraic substitution searches for expressions not explicitly recognized as surface-type operands:

Step 195: Algebraic Substitution

```
> algsubs(x*y=1,expr);
```
Substitute 1 *for each occurrence of* xy *inside* $xy^2 \sin(xt)$.

$$\sin(xt) y$$

algsubs *found that* $xy^2 \sin(xt) = (xy)y \sin(xt)$.

Maple provides means for different ways to perform substitutions. For instance, try **simplify(expr,{x*y=1})** for Step 193. Consult **?simplify[siderel]** for more information.

PRACTICE!

1. Restart Maple and the expression $y = mx + b$ to the name *EQN*.
2. Substitute the values $x = 1$, $m = 2$, and $b = 0$ into *EQN*.
3. What output does the input **subs(x=0,sin(x))** produce? Why is the answer not 0?
4. Substitute $ab = c$ into *abc*. Will entering **subs(a*b=c,a*b*c)** produce useful results? Why or why not? What command should you enter instead?

8.3   EVALUATION

Maple supplies tools for **evaluation** when you need to fully evaluate the result of substitution, which **subs** does not do. Evaluation is a process that breaks expressions into subexpressions, replaces variables with assigned values, and then computes the result. Although Maple uses full evaluation for most expressions, some types only partially evaluate (see **?last_name_eval**).

Use the functions introduced in this section to force Maple into evaluating unevaluated expressions.

8.3.1   Basic Evaluation

Consider the basic substitution statement, **subs(x=Pi/3,sin(x))**, which yields $\sin(\frac{1}{3}\pi)$. If you prefer a fully evaluated result, force full evaluation with **eval(expr)**:

Step 196: Force Evaluation

> **expr := subs(x=Pi/3,sin(x)):**   *Substitute $x = \frac{\pi}{3}$ into $\sin x$. Maple produces $\sin(\frac{1}{3}\pi)$.*

> **eval(expr);**   *Fully evaluate **expr**.*

$$\frac{1}{2}\sqrt{3}$$

Maple evaluated the expression.

For more information, consult **?eval**. Maple offers many functions for evaluation. For instance, new Maple users might prefer **value**. For other evaluation functions, consult Table 8.1 and Mathematics...Evaluation....

8.3.2   Numerical Evaluation

Enter **evalf(expr, n)** to numerically evaluate and report **expr** using **n** digits:

TABLE 8.1   Common Evaluation Functions

Function	Description	Example
eval	evaluate expression	eval(expr)
evalb	evaluate boolean	evalb(1<2)
evalc	evaluate complex	evalc((a+b*I)*(c+d*I))
evalf	evaluate floating-point	evalf(Pi/2,3)
evaln	evaluate to a name	Name:=evaln(Name)
evalr	evaluate range	evalr(INTERVAL(1..2)+INTERVAL(3..4))

Step197: Numerical Evaluation

> **evalf(sin(Pi/3),3);**   *Numerically evaluate $\sin(\frac{1}{3}\pi)$ and report in three digits.*

$$0.865$$ *Maple produced a floating-point result.*

Recall that entering decimal values into expressions typically forces floating-point evaluation.

8.3.3   Substitution and Evaluation

To substitute a new expression for an old one and evaluate the resulting expression, enter **eval(expr, old=new)**. Maple substitutes expression **new** for each occurrence of expression **old**. Then, Maple evaluates the revised expression:

Step198: Evaluation

> **eval(sin(x),x=Pi/3);**   *Substitute $x = \frac{\pi}{3}$ into $\sin x$.*

$$\frac{1}{2}\sqrt{3}$$ *Confirm by entering **sin(Pi/3)**.*

However, because **subs** cannot assign a local variable, neither can **eval**:

Step 199: Evaluation Never Assigns!

> **x;**   *Check the value of x.*

$$x$$ *Entering **eval(sin(x),x=Pi/3)** did not assign x.*

Note that **eval** has difficulty with algebraic substitutions. Unlike **subs** and **algsubs**, however, **eval** has no corresponding "**algeval**."

8.3.4   Multiple Evaluations

As with **subs**, you may refer to **old=new** as **eqn**. Hence, you should enter **{eqns}** instead of **eqn** for multiple equations.

PRACTICE!

5. Evaluate $x^2 + \cos(x)$ for $x = 2$.
6. Evaluate $\dfrac{x^2 - y^2}{x + y}$ for $x = 2$ and $y = 3$.
7. Given $xy = 1$, evaluate $xy^2 \sin(xt)$. Does **eval** work?

8.4   SOLVING

This section presents techniques that solve equations for individual expressions.

Step 200: Exact Solutions

Solving manipulates equations to isolate desired variables. For instance, manually solving an equation $y = mx + b$ for the variable x yields $x = y - b/m$. In Maple, to solve equation **eqn** *exactly* for variable **var**, enter **solve(eqn, var)**:

Step 201: Solve Equation

> **restart:**   *Clear all assignments.*

> **solve(p=k*u,k);**   *Solve p = ku for k.*

$$\frac{p}{u}$$ *Maple reports the solution* $k = \frac{p}{u}$.

As with **subs** and **eval**, **solve** uses local variables during the solution process and does not assign solutions. If you do not believe me, evaluate the variable **k** from Step 201:

Step 202: **solve** Does Not Assign!

> **k;**   *Check the value of k from Step* 201.

$$k$$ *Maple did not assign k.*

To assign a single solution, you have to specify an assignment. For instance, you could directly assign the solution of **solve(p=k*u,k)** to a name, such as *Sol*:

Step 203: Assign a Single Solution

> **Sol:=solve(p=k*u,k);**   *Solve p = ku for k. Assign the the results to* Sol.

$$Sol := \frac{p}{u}$$ Sol *now stores the results.*

Because **solve** uses local variables, the variable **k** does not retain a value after the evaluation of Step 203. Therefore, you could have entered **k:=solve(p=k*u, k)**, which may look a bit odd, but it will indeed work! Also, note that a single equation is also known as a *scalar equation*, as opposed to multiple equations, which are reviewed elsewhere in this text. For more information, consult **?solve** and **?solve[scalar]**. You may find information on related functions in **?SolveTools**, **?eliminate**, **?isolve**, **?isolate**, and **?solvefor**.

8.4.1 Multiple Solutions to Single Equation

Some equations, such as polynomials, yield multiple results. For instance, the equation $x^2 - 4 = 0$ has two solutions: $x = 2$ and $x = -2$. Try finding these solutions using **solve**:

Step 204: Multiple Solutions

> **solve(x^2-4=0,x);**   *Solve* $x^2 - 4 = 0$ *for x.*

$$2, \; -2$$ *Maple reports multiple solutions as a sequence.*

If you expect or discover multiple solutions, use braces with the syntax **solve({eqn}, {var})** to present clearer solutions:

Step 205: Use Braces for Multiple Solutions

> `solve({x^2-4=0},{x});`   *Solve $x^2-4\ =\ 0$ for x. Supply braces to collect solutions.*

$$\{x\ =\ 2\}, \{x\ =\ -2\}$$ *Specifying {} produces solutions as sets.*

8.4.2   Extracting Solutions

Many equations produce multiple solutions. Use square brackets (`[]`) to extract solutions as you would select expressions from a sequence, list, or set. You should first assign the sequence of solutions to a name, like **Sols**:

Step 206: Assign Solutions

> `Sols := solve({x^2-4=0},{x});`   *Solve $x^2-4\ =\ 0$ for x. Assign the sequence of solutions to Sol.*

$$Sols\ :=\ \{x = 2\}, \{x = -2\}$$ *Sols is a sequence of sets.*

Because **Sols** holds a sequence in the form Sol_1, Sol_2 you can select individual solutions by using indices:

Step 207: Extracting Solutions

> `Sols[1];`   *Extract the first solution from Sols.*

$$\{x = 2\}$$ *This expression is the first solution.*

To be even fancier, you could use **eval** to extract the numerical solution for x. How? Note the following:

- Each solution, Sol_1 and Sol_2, forms an equation as **var=val**.
- Entering **eval(expr, var=val)** replaces **var** with **val** inside **expr**.

Therefore, given **name** inside a solution, you can enter **eval(name, name=val)** to extract the value of **name**:

Step 208: Extracting Values from Solutions

> `eval(x,Sols[1]);`   *Substitute $x = 2$ into the expression x. Evaluate the result.*

$$2$$ *$Sols_1$ is {$x = 2$}. Maple extracted the value 2.*

Also, try **lhs(Sols[1])** and **rhs(Sols[1])** to extract the name x and value 2.

8.4.3   Related Commands

Consult Mathematics...Finding Roots, Factorization, and Solving... and **?solve** for numerous related functions. Example worksheets that demonstrate many of **solve**'s techniques are contained in **?examples[solve]**.

PRACTICE!

8. Solve the equation $\sin(x) = \frac{1}{2}\sqrt{3}$ for x.

9. Solve the quadratic equation $ax^2 + bx + c = 0$ for x.

10. Extract the first and second solutions for the quadratic equation. Assign the solutions to $x1$ and $x2$, respectively.

11. Solve $\sin(x) = 0$ for x. Produce all possible values of x. Hint: Consult **?solve** for the environment variable **_EnvAllSolutions**.

8.4.4 Numerical Solutions

Not all equations have exact answers. You may find approximate results with **numerical solutions**. For example, the "floating-point solve" function **fsolve(eqn, var)** finds numerical results relatively quickly:

Step 209: Numerically Solve Equations

```
> fsolve(x^2=2},x);
```
Numerically solve $x^2 = 2$ for x.

$$-1.414213562, 1.414213562$$
Maple reports answers as floats.

You may supply additional options to control **fsolve**'s behavior. For instance, to search for solutions within a specific interval, enter **fsolve(eqn, var, range)**:

Step 210: Find Solution within a Specified Interval

```
> fsolve(tan(x)=x,x,Pi..2*Pi);
```
Numerically solve $\tan(x) = x$ for x within $\pi \le x \le 2\pi$.

$$4.493409458$$
Maple found an answer for the given interval.

Another method involves entering numbers as floats inside **solve**, as discussed in **?solve[float]**. For related functions, consult **?Digits**, **?preferences**, **?evalf**, and **?fnormal**.

PRACTICE!

12. Solve $x^x = 2$ for a numerical value of x. Use both **fsolve** and **solve**.
13. Solve $\sin(x) = 0$ given that $10 \le x \le 20$.
14. Solve $\sin(x) = 0$. Avoid $x = 0$ and $x = \pm\pi$. Hint: Consult **?fsolve**.

8.5 VERIFICATION

Computer programs are not perfect, so always check your work! This section discusses how you can apply Maple functions for solution **verification**, or how you can check your work.

8.5.1 Checking Results

Substituting solutions back into equations should produce **identities** of the form LHS=RHS. For instance, do the following:

- Substitute $\dfrac{y - b}{m}$ for x inside $y = mx + b$.

- Rearrange the equation that you produce, which is $y = m\left(\dfrac{y-b}{m}\right) + b$.
- Eliminate common terms, which produces the identity $y = y$.

Thus, you have verified that $x = \dfrac{y-b}{m}$ is a valid solution $y = mx + b$.

Using Maple, you can enter **eval(eqn, sol)** to verify that **sol** solves **eqn**. Recall that **sol** has the form **var=val**. For instance, try confirming that Maple's solutions for $x^2 - 4 = 0$ are indeed valid:

Step 211: Verify Solutions

```
> Eqn := x^2-4=0:
```
Assign the equation $x^2-4 = 0$ to Eqn.

```
> Sols := solve({Eqn},{x});
```
Solve $x^2-4 = 0$ for x. Assign the solutions to Eqn.

$$Sols := \{x = 2\}, \ \{x = -2\}$$
Maple found two solutions.

```
> eval(Eqn,Sols[1]), eval(Eqn,Sols[2]);
```
Substitute solutions into Eqn.

$$0 = 0, 0 = 0$$
Maple produces identities for both solutions.

8.5.2 Multiple Substitutions

Try **map(subs, [sols], eqn)** for multiple substitutions. Maple expands this command as the sequence **subs(sols[1], eqn),subs(sols[2],eqn),...,subs (sols[n], eqn)** for **n** solutions. You must use **[sols]** instead of **sols** because **map** requires a list. So, using the information in Step 211, you would enter the following:

Step 212: Verify All Solutions

```
> map (subs, [Sols], Eqn);
```
Substitute each solution into Eqn. Use lists with map!

$$[0 = 0, 0 = 0]$$
The solutions produce identities.

Remember that **subs** does not evaluate expressions! Extract and evaluate individual results when **subs** does not produce identities. Consult **?map** for more information.

PRACTICE!

15. Solve $x^2 + 4x + 3 = 0$ for x. Verify each solution individually.
16. Verify all solutions in the previous problem by using **map**.

8.6 MISCELLANEOUS

This section introduces further features of **solve** and related commands.

8.6.1 RootOf

Solving for roots sometimes produces odd looking results, especially when dealing with fourth-order or higher polynomials. Because solutions may include complex results, Maple will prefer to "hide" the full results with a placeholder called **RootOf**. Often, the higher order expressions will also yield multiple solutions, so Maple outputs each solution in the syntax **RootOf(*expr, index*)**, which indicates an actual root of ***expr***. For instance, investigate the roots of $x^4 + x - 3$ by using **solve**:

Step 213: RootOf

> **x := 'x':**   *Clear any value stored in x.*

> **Sols := solve(x^4+x−3=0,x);**   *Find the roots of x^4+x-3.*

Maple produces four possible solutions, which are "hidden" by using **RootOf:**

Sols := RootOf(_Z^4+_Z-3,*index* = 1), RootOf(_Z^4+_Z-3,*index*=2),
   RootOf(_Z^4+_Z-3,*index*=3), RootOf(_Z^4+_Z-3,*index*=4)

Here, Maple found four solutions and numbered them. In case you are worrying about the _Zs, Maple distinguishes **RootOf** solutions from the original equation by replacing your independent variable with a default _Z variable. You do not have to use the _Zs, as you will see shortly.

   To find explicit results, you have a few choices. I suggest that you try numerical solutions by evaluating the solution set with **evalf**:

Step 214: Explicit Solutions (Numerical Results)

> **evalf(Sols,3);**   *Find the roots of x^4+x-3.*

Maple produces four possible solutions, which are "hidden" because of **RootOf:**
   1.16, 0.144+1.32I, -1.45, 0.144-1.32I

Maple solved for all the _Zs in each of the four solutions. If you need exact results, you should try **allvalues**, which will produce a rather large set of symbolic solutions. For example, to find the exact expression for the first solution, do the following:

Step 215: Find All Roots

> **allvalues(Sols[1]);**   **allvalues** *symbolically evaluates all* **RootOf** *values.*

You will see very a lengthy result!   *Use a colon next time.*

Maple has extensive help on the functions described in this section. You should read about the details in **?solve**, **?RootOf**, and **?allvalues**. For related functions and useful commands, investigate **?RealDomain**, **?evalf**, **?convert[RootOf]**, **?convert[radical]**, and **?type[RootOf]**.

8.6.2   Labeling

When solving high-degree polynomial equations, Maple tends to generate rather long solutions. Maple can break up long output into ***labels***, which are names that start with a percent sign **(%)**. To activate this feature, select File→Preferences... and select the I/O Display tab. For your output, choose Typeset Notation. If you go back and re-enter your input, Maple will report names, like **%1** and **%2**, which you may use as input. Investigate **labeling** and **labelwidth** options inside the **?interface** Help window for more details.

8.6.3 Polynomials

Find polynomial roots with **solve** or **roots** by entering **solve(poly=0,name)** or **roots(poly)**. (Polynomials are discussed in Chapter 5.)

8.6.4 Lost Solutions

If **solve** fails to find a solution, Maple produces no output. If you suspect **solve** missed some solutions, check if Maple assigned the Boolean value *true* to a global variable called **_SolutionsMayBeLost**. Consult **?solve** for more details.

8.6.5 Inequalities

Inequality expressions contain **<**, **<=**, **>**, and **>=**, and are called *relations*. The **solve** function can solve relations of a single inequality:

Step 216: Solve an Inequality

> **solve(2*abs(x)>1,{x});**   *Solve* $2|x|>1$.

$$\left\{\frac{1}{2} < x\right\}, \left\{x < \frac{-1}{2}\right\}$$ *Maple automatically converts inequalities*

into "<" form.

Consult **?solve[ineq]**. Also, consult **?simplex** and **?plots[inequal]** for solving multiple inequality relations.

8.6.6 Extrema

Extrema are extreme values of an expression—either the most negative or the most positive. You may find candidates for extrema of an expression by entering **minimize(expr)** and **maximize(expr)**:

Step 217: Find Minimum and Maximum Values

> **f := sin(x):**   *Assign a function.*

> **minimize(f), maximize(f);**   *Find possible extrema of* $\sin x$.

$$-1, 1$$ *You should further investigate these values.*

For more information, consult **?minimize**, **?extrema**, and **?min**.

PRACTICE!

17. Evaluate RootOf($\_Z^2 + 1$). Find all possible values.
18. Find the roots of $x^5 + x + 1$.
19. Solve $x^2 > 1$ for x.
20. What are the extrema of $\cos(x)$? What are the extrema of $\tan(x)$?

8.7   APPLICATION: SHEET PILES

This section solves a problem involving soil mechanics that requires the solution of a fourth-order polynomial equation.

8.7.1 Background

Landscaping and construction often require structures to shore up and block portions of the ground to keep it from collapsing. Very often, temporary piles help protect workers

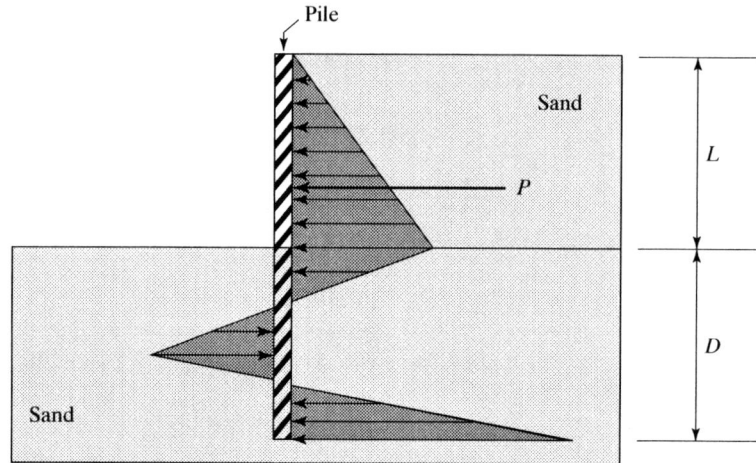

Figure 8.1   Sheet Pile

on construction sites. Figure 8.1 illustrates a sheet pile driven into sandy soil. The weight of the ground creates a force P that pushes the pile to left and causes different pressure distributions to the left and right of the pile.

Balancing these pressure distributions according to the soil properties yields the proper depth to which one should drive the pile into the soil. Given unit weight γ (force/volume), find the necessary pile depth D from the following equation:

$$D^4 - \frac{8P}{\gamma(K_p - K_a)}D^2 - \frac{12PL}{\gamma(K_p - K_a)}D - \left(\frac{2P}{\gamma(K_p - K_a)}\right)^2 = 0 \qquad (8.1)$$

The coefficients

$$K_a = \tan^2\left(45° - \frac{\phi}{2}\right) \qquad (8.2)$$

and

$$K_p = \tan^2\left(45° + \frac{\phi}{2}\right) \qquad (8.3)$$

help determine the pressure that the soil places on the pile. The angle ϕ represents the soil's friction.

8.7.2   Problem
Given $\gamma = 18\ \text{kN/m}^3$, $\phi = 30°$, $L = 3$ m, and $P = 30$ kN/m, determine the necessary depth of the sheet pile shown in Figure 8.1.

8.7.3   Methodology
First, state and assign pertinent variables:

Step 218: Sheet Piles—Initialize Maple

```
> restart:
```
Restart your Maple session.

```
> with(Units[Natural]):
```
Activate Maple's unit capabilities.

Ignore the lengthy warning that Maple gives.

Step 219: Sheet Piles—Restate and Separate

```
> unprotect(gamma): gamma := 18*kN/m^3:
```
Maple predefines gamma.
```
> phi := 30*degrees:
```
Assign the flow height.
```
> theta := 45*degrees:
```
Assign the notch angle but use radians!
```
> L := 3*m: P := 30*kN/m:
```
Assign length and force.

To simplify data entry, assign the model for Eq. (8.1) first:

Step 220: Sheet Piles—Parameters

```
> c1 := 8*P/(gamma*(Kp-Ka)):
```
$$c1 = \frac{8P}{\gamma(K_p - K_a)}$$
```
> c2 := 12*P*L/(gamma*(Kp-Ka)):
```
$$c2 = \frac{12PL}{\gamma(K_p - K_a)}$$
```
> c3 := 2*P/(gamma*(Kp-Ka)):
```
$$c3 = \frac{2P}{\gamma(K_p - K_a)}$$
```
> Ka := tan(theta-phi)^2;
```
$$K_a = \tan^2\left(45° - \frac{\phi}{2}\right)$$
```
> Kp := tan(theta+phi)^2;
```
$$K_p = \tan^2\left(45° + \frac{\phi}{2}\right)$$

Now, you ought to be able to form the model, but you will encounter a slight problem:

Step 221: Sheet Piles—Model

```
> Eqn := depth^4 - c1*depth^2 - c2*depth - c3^2 = 0;
```
Assign the model.
```
Error, (in +) the units '1' and 'm^2' have incompatible
dimensions
```

What happened? Maple does not know that depth has units of length in terms of meters. So, you might want to try the following trick:

Step 222: Sheet Piles—Model (Revisited)

```
> dp := depth*[m]:
```
Assign a temporary variable that will have units of length.
```
> Eqn := dp^4 - c1*dp^2 - c2*dp - c3^2 = 0;
```
Assign the model again.

Maple evaluates the model with units now. Note how the units balance for each term:

$$Eqn := \left(depth^4 - 5depth^2 - \frac{45}{2}depth - \frac{25}{16}\right)[m^4]$$

Now, solve the fourth-order polynomial equation *Eqn* for the pile depth:

Step 223: Sheet Piles—Solve

> `Sols := solve(Eqn,depth);`   *Solve the model for depth.*

Maple produces four results in terms of **RootOf** *which are not shown to conserve space.*

Remember that Maple often produces results in terms of **RootOf** when solving roots for fourth-order or higher polynomials. Since a contractor will not understand a **RootOf** answer on a specification, you should find a numerical answer:

Step 224: Sheet Piles—Report

> `evalf(Sols,3);`   *Find a numerical solution.*

$$3.42, \ -1.68 + 1.92I, \ \ -0.0750, \ -1.68 - 1.92I$$ *Maple produces unwanted results!*

8.7.4   Solution

Fortunately, you can readily identify the real answer of 3.42 m because the pile depth cannot realistically be imaginary or negative, let alone both![1] Actually, Maple can assist you with finding the real-valued result by using the **RealDomain** package:

Step 225: Sheet Piles—Solution

> `Sols := evalf(RealDomain[solve](Eqn,depth),3)*m;`

$$3.42[\mathrm{m}]$$ *Maple found only the real-valued solution.*

KEY TERMS

algebraic substitution	evaluation	extrema
global variables	identities	labels
numerical solutions	scalar equation	solving
syntactic substitution	verification	

SUMMARY

- Use syntactic substitution with **subs** when all you need to do is substitute a variable for another.
- Use algebraic substitution with **algsubs** when subs fails.
- Use **eval(*expr, eqn*)** when you need to substitute an expression and evaluate the result.
- Use **eval(*expr*)** when you need to force Maple into evaluating *expr*.
- Use **solve** to find exact solutions to equations.
- Use **fsolve** to find numerical solutions to equations.
- Use **evalf** or **allvalues** if Maple produces results in terms of **RootOf**.
- **subs**, **eval**, and **solve** use local variables during evaluation and, consequently, do not assign results.
- Check your results to ensure that Maple found the correct expressions!

---

[1] What happened to the [*m*] in the output? Maple divides out the units because they are a common factor in the equation. For instance, try **with(Units[Natural]): solve(x*m−4*m=0,x);**. Maple will report 4 without units.

Problems

1. Explain the differences between Maple's substitution, evaluation, and solving procedures.

2. When should you use syntactic substitution as opposed to an algebraic substitution?

3. Describe cases in which you would use **eval** in place of **subs**.

4. Perform the following substitutions using the **subs** command:

    (a) Find $f(2)$ given $f(x) = 1 + x$.

    (b) Find $f(2)$ given $f(x) = \dfrac{x + c}{2}$.

    (c) Find $f(1, 2)$ given $f(x, y) = x - y^2$.

    (d) Find $f(0)$ given $f(x) = \dfrac{\sin(x)}{x}$.

    (e) Find $f(2)$ and $g(2)$ given $f(x) = 1 + x^2$ and $g(x) = \sqrt{f(x)}$.

    (f) Find $f(2)$ and $g(2)$ given $f(x) = 1 + x^2$ and $g(x) = f(x + 1)$.

5. Repeat Problem 4 using the **eval** command instead of **subs**.

6. Replace the subexpression $\frac{a}{c}$ with x inside the expression $\frac{ab}{c}$ using a substitution. Hint: You should try **subs**, but that function will fail. Use **algsubs** instead.

7. Find the numerical value of π to 5 decimal places without changing the value of **Digits**.

8. Solve the equations that follow for exact solutions when possible. Otherwise, find floating-point answers. Hints: In general, you should use **solve**. You might wish to find numerical solutions **fsolve** when **solve** fails. Note also that some equations might generate complex solutions.

    (a) Solve $y = mx + b$ for x.

    (b) Solve $y = 5x - 1$ for x.

    (c) Solve $11x^2 - 0.17x = -5.4$ for x.

    (d) Solve $\sin\left(x + \dfrac{\pi}{3}\right) = 1$ for x.

    (e) Solve $\ln\left(\dfrac{x}{x + y}\right) + x = 1$ for y.

    (f) Solve $\sqrt{x + 2\sqrt{x}} - \sin(x + 1) = 0$ for x. Produce a complex result.

9. How accurate is **fsolve** when **Digits** has its default value? Explain your result using the equation $x^2 = 2$.

10. Verify the solutions for Problem 8. Hints: Try to substitute the solutions back into the original equations. You might need to simplify or expand various subexpressions.

11. Solve $x^5 - 3x^3 + x - 1 = 0$ for x. Hints: Applying **solve** yields two solutions. Use **allvalues** on the more complicated of the two. Then, use **evalf**.

12. Find a general formula for solving the general third-order polynomial,

$$a_0 + a_1 x + a_2 x^2 + a_3 x^3 = 0, \tag{8.4}$$

for x. Hints: Use **solve**, and supply curly braces for the variable x. You can use either arbitrary constants or the constants a_i that are shown in Equation (8.4). For instance, to output a_2, you would enter **a[2]**. Also, you should consider activating Maple's labeling feature, as discussed in **?labeling**.

13. Repeat the problem in Section 8.7 given $\phi = 25°$ and $P = 35$ kN. Verify your results using Maple.

14. Given $\gamma = 18$ kN/, $\phi = 30°, \theta = 45°, L = 3$ m, and $P = 24$ kN/m, determine the necessary depth of the sheet pile shown in Figure 8.1. Is this value of depth greater or less than the value calculated in Section 8.7? Explain why.

15. Thermodynamics provides many equations for relating the physical properties of substances. Given volume v, temperature T, pressure P, gas constant R, and arbitrary constant C, the equation

$$v = \frac{RT}{P} - \frac{C}{T^3} \tag{8.5}$$

defines the equation of *state* for a gas. Find the temperature T in terms of the other variables. Hint: You should consider activating Maple's labeling feature.

16. What interest rate i will convert a current payment $P = \$12{,}000$ into an annual cost $A = \$2400$ for a 10-year period? Hint: Refer to Section 3.12.

17. Repeat Problem 11 in Chapter 6, but use **solve** to find values of τ from $\tau^3 - A\tau^2 + B\tau - C = 0$.

18. Snow fences help to prevent snow drifts from accumulating on roads. Given fence length L_f, fence height H, storage capacity Q_c, and storage capacity for an infinitely long fence $Q_{c,\,inf}$, engineers have developed the following empirical equation:

$$\frac{Q_c}{Q_{c,\,inf}} = 0.288 + 0.039\left(\frac{L_f}{H}\right) - 0.0009\left(\frac{L_f}{H}\right)^2 + \frac{\left(\frac{L_f}{H}\right)^3}{133333} \text{ for } 5 \le \frac{L_f}{H} \le 50. \tag{8.6}$$

Assuming a factor $\frac{Q_c}{Q_{c,\,inf}} = 0.8$ and a fence height $H = 6\,ft$, solve for the required length L_f.

19. The "loudness" of sound L (decibels, or dB) can be estimated using a logarithmic intensity ratio:

$$L = 10\log_{10}\left(\frac{I}{I_0}\right). \tag{8.7}$$

I is the intensity (W/m², or Watts per square meter). The threshold of human hearing is defined as the intensity $I_0 = 1 \times 10^{-12}$ W/m².

(a) In general, a sound that has intensity $I = 1$ reaches the threshold of pain in humans. To what level of loudness L does this intensity correspond? Hints: Evaluate the equation for L given $I = 1$ and $I_0 = 1 \times 10^{-12}$. Note that you are using a base-10 logarithm. Use a different name than I (why?). You should compute 120 dB.

(b) The heavy metal band MANOWAR claimed to achieve a world record loudness of 130 dB at a concert in 1994. What value of intensity I did the band reach? Based on your results and the information in Problem 19a, were the attendees in pain?

9

Systems of Equations

9.1 APPLICATION: SIMULTANEOUS EQUATIONS

I want to motivate this entire chapter with a simple, but useful, and elegant example that has applications in all forms of engineering. This section develops the model and discusses general mathematical concerns to solve it. Later chapters will demonstrate a variety of approaches for the solution.

OBJECTIVES

After reading this chapter, you should be able to

- Describe and model systems of equations
- Solve systems of equations by hand
- Solve systems of equations with Maple plots
- Introduce linear algebra with vectors and matrices
- Solve systems of equations with Maple's linear algebra functions

9.1.1 Complexity of Models

In general, increasing the complexity of a model introduces more variables. The universe has very many factors that influence another multitude of systems. To model a system with many influences, the developer of the model introduces a variety of variables that interact with each other. This section introduces the concept of such a multivariable model and the methods used to solve it. Before solving a model, however, you will first need to develop it. An example of the development and solution processes is shown in this section.

9.1.2 Building a Model

A device composed of two elastic bars capped with rigid plates is shown in Figure 9.1. Loads are statically applied on both plate ① and plate ②. We will denote the connections at ① and ② as *nodes*.

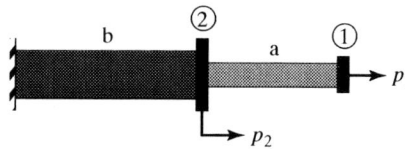

Figure 9.1 Example Device

You can model elastic bars as springs, as shown in Figure 9.2. Assume that Hooke's law ($p = ku$) governs spring behavior. Thus, bar a has force p_a and displacement u_a. Similarly, bar b has force p_b and displacement u_b.

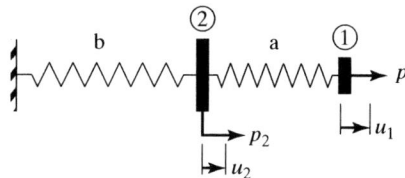

Figure 9.2 Model of Device

After the loads are slowly applied, the bars deform and reach a new resting, or *equilibrium*, position. The applied loads (p_1 and p_2) and internal forces of the bars (p_a and p_b) must balance according to equilibrium. Assume no twisting or rotation of the plates occurs. (See Figure 9.3.)

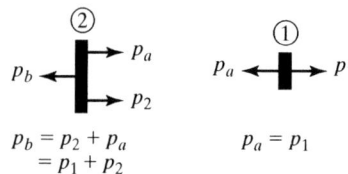

$$p_b = p_2 + p_a \qquad p_a = p_1$$
$$= p_1 + p_2$$

Figure 9.3 Free Body Diagrams

From Hooke's Law, illustrated in Figure 9.4, you can relate each spring's internal force to its relative displacement, the measure of how much each spring stretches. Bars a and b have relative displacements u_a and u_b, respectively.

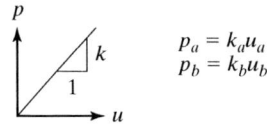

$$p_a = k_a u_a$$
$$p_b = k_b u_b$$

Figure 9.4 Hooke's Law

Inspect Figure 9.5 to determine relationships for displacements u_a and u_b. The applied forces stretch spring b an amount u_b. Since the other end of b is fixed, node ② moves the same amount as the total stretch in spring a, resulting in $u_b = u_2$. For u_1, the displacement at node ② and the amount of stretch at node ① caused by the applied force both contribute. Therefore, $u_1 = u_2 + u_a$, and as a result, $u_a = u_1 - u_2$.

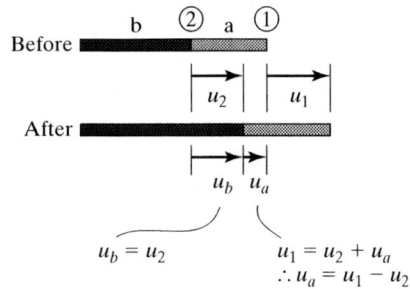

$$u_b = u_2$$
$$u_1 = u_2 + u_a$$
$$\therefore u_a = u_1 - u_2$$

Figure 9.5 Relative Displacements

To ease your analysis, you should express the equations in terms of nodal values, as shown in Figure 9.6. Combine equilibrium, Hooke's Law, and displacement relations into two equations in terms of p, k, and u.

$$p_1 = p_a$$
$$= k_a u_a = k_a(u_1 - u_2)$$
$$p_2 = p_b - p_1$$
$$= k_b u_2 - k_a(u_1 - u_2)$$
$$= -k_a u_1 + (k_a + k_b)u_2$$

Figure 9.6 Combine Equations

9.1.3 Systems of Equations

To fit a more standard appearance, which you might see later if you continue your studies, you should rearrange the equations of Figure 9.6 into the following **system of equations**:

$$k_a u_1 - k_a u_2 = p_1 \tag{9.1}$$

$$-k_a u_1 + (k_a + k_b)u_2 = p_2 \tag{9.2}$$

A system of equations collects simultaneous equations with common unknown variables, also called *unknowns* or *indeterminates*. The system in Eqs. (9.1) and (9.2) has two unknowns, u_1 and u_2. Assume that all other variables have predetermined values.

① Substitute values into Eqs. 9.1 and 9.2.	② Reduce the second equation by adding the first.	③ Divide equations by the first, or leading, coefficient.	④ Backsubstitute results into the first equation.
$2u_1 - 2u_2 = 10$ $-2u_1 + 5u_2 = 20$	$2u_1 - 2u_2 = 10$ $3u_2 = 30$	$u_1 - u_2 = 5$ $u_2 = 10$	$u_1 = 15$ $u_2 = 10$

Figure 9.7   Gaussian Elimination

A ***linear system*** of equations is a system of equations that contains only first-order equations in the form $a_0 = a_1 x_1 + \dots + a_n x_n$. Both Eqs. (9.1) and (9.2) contain terms with powers no greater than unity, and thus, the equations constitute a linear system. On the other hand, a ***nonlinear system*** of equations contains terms with powers higher than unity.

9.1.4   Gaussian Elimination

To solve for the unknowns, u_1 and u_2, follow the steps, as shown in Figure 9.7. First, assume the values $k_a = 2, k_b = 3, p_1 = 10$, and $p_2 = 20$ in Step ①. Next, you will apply ***gaussian elimination***, as demonstrated by Steps ②, ③, and ④. By adjusting coefficients, you can eliminate common terms. After dividing out leading coefficients, you can solve for unknowns by back substitution. For a more detailed manual approach, see Table 9.3.

9.1.5   Dependency

What kind of guarantee do you have that you will find a unique set of solutions when solving a linear system? You need a ***linearly independent*** system of equations, in which the following conditions hold:

- The number of equations matches the number of unknowns.
- No equation duplicates, or is a multiple of, another equation in the system.

A linearly independent system produces only one unique solution for each unknown. You might encounter systems with duplicates of equations, such as the system $x + y = 1$, and $2x + 2y = 2$. Such a system produces an infinite number of x and y solutions and is called ***linearly dependent***.

9.1.6   Solution . . . ?

Throughout this chapter, you will learn a variety of techniques to solve the problem presented in the section. Maple, in particular, is extremely useful for solving such problems.

9.2   GENERAL EQUATION SOLVING

This section demonstrates the application of **solve** to linear and nonlinear systems of equations.

9.2.1   Multiple Equations

To solve a system of equations, enter **solve({eqns},{vars})**, where

- **{eqns}** is the set of simultaneous equations in terms of **vars**.
- **{vars}** is the set of unknown variables that you wish to find.

Note that the system may be linear or nonlinear. For instance, try solving the spring model from Section 9.1:

Step 226: Solve Linear System

```
> restart:
```
Begin a new Maple session.

```
> eqn1 := 2*u1 - 2*u2 = 10:
> eqn2 := -2*u1 + 5*u2 = 20:
```
Assign $2u_1 - 2u_2 = 10$ *to eqn1.*
Assign $-2u_1 + 5u_2 = 20$ *to eqn2.*

```
> sols := solve({eqn1,eqn2},{u1,u2});
        sols := {u2 = 10, u1 = 15}
```
Solve the system of equations.
Maple produces both answers.

For more information, consult **?solve[system]** and **?solve[linear]**.

9.2.2 Verifying Solutions

Remember to check your work. You may use **subs** to substitute your solutions back into the original equations. If everything is OK, Maple will produce a set of identities:

Step 227: Checking Solutions

```
> subs(sols,{eqn1,eqn2});
        {10 = 10, 20 = 20}
```
Substitute solutions into original equations.
Identities indicate correct results.

9.2.3 Assigning Solutions

The **solve** function never assigns the unknowns for which you solve. In general, you may assign unknowns with **assign(expr)**, where **expr** is a list or set that contains the sequence of solutions, **var1=val1, var2=val2,...**. For instance, enter **assign(sols)** to assign $u1$ and $u2$:

Step 228: Assigning Solutions

```
> assign(sols);
```
sols is {u1 = 15, u2 = 10}. Assign 15 to u1. Assign 10 to u2.

```
> u1,u2;
            15, 10
```
Check the values of u1 and u2.
Maple correctly assigned the names.

Consult **?assign** for more information.

9.2.4 Nonlinear Equations

Use **solve** for nonlinear systems. You can mix linear and nonlinear forms as well, though mixed forms are still considered nonlinear systems:

Step 229: Solve Nonlinear Equations

```
> A := x^2+y^2=1:
```
Assign a circle $x^2 + y^2 = 1$ *to A.*

```
> B := x+y=1:
```
Assign a line $x + y = 1$ *to B.*

```
> solve({A,B},{x,y});
    {x = 1, y = 0},  {x = 0, y = 1}
```
Solve equations for common variables.
The line crosses the circle in two places!

9.2.5 Numerical Solutions

You can enter **fsolve({eqns},{vars},options)** to obtain floating-point results of linear and nonlinear systems of equations.

9.3 GRAPHICAL SOLUTION

Both Eqs. (9.1) and (9.2) are mutually dependent equations of u_1 and u_2. Recall that **implicitplot** graphs functions with mutually dependent variables. Thus, you can enter **implicitplot({eqns},h,v)** to display a graphical solution of **eqns**:

PRACTICE!

1. Solve the system $2x + 3y + z = 0$, $2y + z = -1$, and $x + z = 2$.
2. Verify and assign the solutions in the preceding problem. Show values of x, y, and z.
3. Solve the system $x + y = \pi$ and $\sin(x) = y$ for x and y.
4. Can you "solve" the system $x + y = 1$ and $2x + 2y = 2$? Why, or why not?

Step 230: Graphical Solution—Load Packages

```
> restart:
```
Begin new Maple session.

```
> with(plots):
```
You will need **implicitplot**, **textplot**, *and* **display**.
Ignore the warning message.

```
> with(plottools):
```
You will need **circle**.
Ignore the warning message.

Step 231: Graphical Solution—Equations and Plot Options

```
> eqns := {2*u1-2*u2=10,-2*u1+5*u2-20}:
```
Use the spring equations from Step 226.

```
> opts: = thickness=3, tickmarks=[2,2],
        font=[HELVETICA,BOLD,9], scaling=CONSTRAINED:
```
Set basic plotting options.

Step 232: Graphical Solution—Plot Structures

Store an implicitplot structure of the two spring equations:
```
> IP := implicitplot(eqns,u1=0..25,u2=0..25):
```

Store a textplot structure of labels that will you will superimpose on the implicitplot:
```
> TP := textplot ([15,12,"SOLUTION"],align={ABOVE,LEFT},opts):
```

Plot a circle that surrounds the solution at u1 = 15 and u2 = 10:
```
> CP := circle([15,10],1,opts):
```

Step 233: Graphical Solution—Display Plots

Superimpose all three plots for a really interesting depiction of the graphical solution to the system of equations:

```
> display({IP,TP,CP});
```

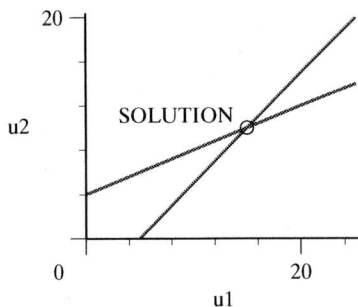

The circle indicates the solution to both equations.

Actually, you technically need to know the solution before being able to know where to plot the circle. So, you could plot both equations without the circle and then inspect the axes, which represent each unknown variable in the equations. The lines cross at the common solution to both equations. Only one common point indicates a unique solution, which you may then approximate as the center of the circle.

PRACTICE!

5. Graphically solve $3x + 2y = -4$ and $-x + 3y = 5$.
6. Can you graphically solve $x^2 + y^2 = 1$ and $y = 2$? Do the equations intersect?

9.4   LINEAR ALGEBRA IN MAPLE

When solving linear systems of equations, data entry becomes tedious. An entire field of study called *linear algebra* is devoted to the study of these equations. Linear algebra uses special structures to collect the multitudes of data.

9.4.1   Library Packages

After assembling the systems of equations, there are many choices for solution techniques, many of which Maple implements in two library packages:

- **linalg** is an older package that suits smaller problems. See **?linalg**.
- **LinearAlgebra** is a newer, more efficient package that readily handles larger problems. See **?LinearAlgebra** or **?LA**.

Maple describes both packages in more detail in Mathematics…Linear Algebra…Linear Algebra Computations (**?linalgebragen**). I will review the **LinearAlgebra** package because it is more robust and a bit easier to understand, especially with Maple's extensive help. I suggest that you remember just **?LAOverview**, which summarizes many of the other overviews on different facets of linear algebra in Maple.

9.4.2 Accessing Maple's Linear Algebra Package

To access the **LinearAlgebra** functions, I suggest that you use Maple's short-form access for this chapter:

Step 234: Accessing **LinearAlgebra** Package

```
> with(LinearAlgebra):
```

For help on a specific *function* in **LinearAlgebra**, consult **?LinearAlgebra** and **?LinearAlgebra[function]**. Consult Appendix B for more help on Maple's library packages.

9.4.3 Maple's Terminology

You might encounter some confusing terminology in the Maple's on-line help when investigating the **LinearAlgebra** and **linalg** packages. Without getting mired in the nitty-gritty details, here is the "gist" of these terms:

- *rtable*: A Maple *table* is a general collection of data with customized indices, as demonstrated in Appendix D. An *rtable* is essentially a *rectangular* table where you can store a variety and mixture of Maple types. See **?table** and **?rtable** for more information.

- *module*: In Maple, you may write your own functions, which are called *procedures*, when you use Maple as a programming language. A module is essentially a collection of procedures and additional data. See **?module** for more information.

- *constructor*: Maple uses a special procedure called a constructor to "build" a variety of objects, such as floats (**Float**), complex numbers (**Complex**), and data structures (**Vector**, **Matrix**) used in **LinearAlgebra**.

- *last name evaluation*: If you use **linalg**, you will discover that Maple will not completely evaluate your data structures because of their "large" size. So, you will need to force evaluation at the last step to show results. See **?last_name_eval** for more information.

PRACTICE!

7. Demonstrate how to access the functions inside Maple's **LinearAlgebra** package.

8. Explain why the following execution group does not produce the output [1]:
```
[> with(LinearAlgebra): restart: IdentityMatrix(1);
```

9.5 VECTORS

Imagine that you need to solve a system of one million equations. Such situations do arise, especially for detailed analysis of intricate structures and other systems. To help you store the data associated with a system of equations, you may use powerful tools called *vectors*, which are discussed in this section. You should also consult **?aboutdata**.

9.5.1 Vector Definition

A ***vector*** represents a quantity with both direction and magnitude. Common examples include force, velocity, and acceleration. You can also use a vector to represent a solution

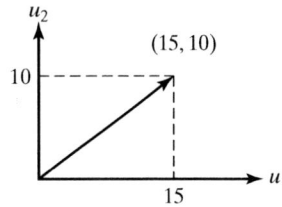

Figure 9.8   Vector Plot

to a linear system. How? For instance, plot (u_1, u_2) values as the coordinate $(15, 10)$, as shown in Figure 9.8:

- Let u_1 and u_2 represent a horizontal and vertical axis on a 2D plot. All coordinates have locations (u_1, u_2).
- Plot the solution $(u_1, u_2) = (15, 10)$ from Section 9.5.
- Draw an arrow from the origin to the solution point $(15, 10)$.

So, why is the quantity plotted in Figure 9.8 a vector? You can compute the following aspects:

- *magnitude*:   Find the length using the Pythagorean theorem $\sqrt{u_1^2 + u_2^2} = \sqrt{15^2 + 10^2}$.
- *direction*:   Orient the vector by drawing an arrow between $(0, 0)$ and $(15, 10)$.

9.5.2   Vector Notation

Let u represent the arrow drawn from $(0, 0)$ to any coordinate (u_1, u_2). The terms u_1 and u_2 are called **vector components**. You write a vector in terms of components in one of the following ways:

- a *row vector*:

$$u = \begin{bmatrix} u_1 & u_2 \end{bmatrix} \tag{9.3}$$

- a *column vector*:

$$u = \begin{bmatrix} u_1 \\ u_2 \end{bmatrix}. \tag{9.4}$$

Which should you choose? For this text, we will stick mostly to column vectors because they often appear in systems of equations. When writing vectors by hand, you should distinguish vectors from other quantities with arrow notation (e.g., $\bar{u}$ or $\vec{u}$). Also, vectors with unitary magnitudes (length of 1 or unity) often are written as *unit vectors* $\hat{u}$. See **?VectorCalculus** for Maple's application of unit vectors for coordinate systems.

9.5.3   Vector Terminology
Maple uses different terms for a mathematical vector:

- *Vector*: a vector created from the **LinearAlgebra** package
- *vector*: a vector created from the **linalg** package

For this text, *vector* refers to the mathematical notion of a vector to avoid confusion. The term **Vector** will match Maple's definition. I do not demonstrate **linalg**'s *vector*.

9.5.4 Vector Syntax

In Maple, there are many ways to enter a **Vector**. I will focus on **Vector(list)** to automatically create a column vector that consists of values from **list**. The elements of **list** become the elements of the vector in the same order. For instance, to store 15 and 10 in a vector, you would do the following:

Step 235: Create Vector

> `u := Vector([15,10]);`

Assign the vector $\begin{bmatrix} u_1 \\ u_2 \end{bmatrix}$ *to u.*

$$u := \begin{bmatrix} 15 \\ 10 \end{bmatrix}$$

If you would like to use a row vector, enter **Vector[row]([10,15])**. If you would like to see the options that Maple uses for a **Vector** called **u**, enter **VectorOptions(u)**. For more information, consult **?Vector** and **?VectorOptions**.

9.5.5 Vector Palette and Shortcut Notation

If you cannot remember the syntax for creating a **Vector**, use the vector palette by selecting <u>V</u>iew→<u>P</u>alette→<u>V</u>ector Palette, as shown in Figure 9.9. If you click any pattern, Maple will show a **Vector** using shortcut notation:

- **<v1, v2, ..., vn>** creates a column vector, because each comma (**,**) starts a new row.
- **<v1|v2|...|vn>** creates a row vector, because each bar (**|**) starts a new column.

For instance, after typing **u:=** and selecting ▦ from the Vector Palette, you would have the following input with placeholders:

Step 236: Create Vector with Shortcut Notation

*Enter **u:=**, then use the Vector Palette. Replace the placeholders with values:*
> `u := <%?, %?>;`

A highlighted placeholder indicates where the next input will go. You may use the Tab key to switch to a new placeholder. Press Enter after you have finished entering your values.

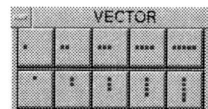

Figure 9.9   Vector Palette

Maple allows you to enter shortcut notation without using the Vector Palette. For instance, if you want to create a row vector, enter the following:

Step 237: Create Row Vector

```
> rowvec := <1|2>;
```
$$rowvec := [1\ 2]$$

Create a row vector.

Each bar (|) starts a new column. To create a column vector, enter the following:

Step 238: Create Column Vector

```
> colvec := <1,2>;
```
$$colvec := \begin{bmatrix} 1 \\ 2 \end{bmatrix}$$

Create a column vector.

Each comma (,) starts a new row. Consult **?vecpalette** and **?MVshortcut** for more information.

9.5.6   Vector Extraction

Use indexed names with the syntax **name[index]** to access individual vector components. Indices start from 1:

Step 239: Extract Vector Components

```
> u[1],u[2];
```
$$15, 10$$

Extract u_1 and u_2.

Vector components $u_1 = 15$ and $u_2 = 10$.

As discussed in **?MVextract**, you may extract multiple elements, as in **u[1..2]**, **u[[1,2]]**, and (believe it or not!) **u[[-2, -1]]**. See **?MVselect** for related help. Consult **?list** and **?name** for general information on extraction and indexed names.

9.5.7   Vector Assignment

Assuming you do not set your **Vector** to option **readonly** (which means, *unchangeable*), you may reassign some, or all, of its components. For instance, suppose you wish to replace the second element of a **Vector**. You would use the same syntax from extraction, though now the index appears on the LHS of the assignment:

Step 240: Assign Vector Components

```
> v := Vector([1,2]):
```
Create a column vector.

```
> v[2] := 72;
```
$$v_2 := 72$$

Change the second element, which is on row 2, to 72.

Note the indexed name v_2.

```
> v;
```
$$\begin{bmatrix} 1 \\ 72 \end{bmatrix}$$

Check the vector to see the changed value.

Yes, the second row did get the new value.

For more information and rules, consult **?MVassignment**.

PRACTICE!

9. Assign $\begin{bmatrix} 11 \\ 21 \end{bmatrix}$ to p.
10. Check p's type.
11. Assign 10 to a and 20 to b and show p.

9.6 MATRICES

To help you store the data associated with a system of equations, you may use powerful tools called *matrices*, which are discussed in this section. You should also consult **?aboutdata**.

9.6.1 Matrix Definition

A vector stores information in one dimension, either horizontally or vertically. What should you do to model data that is tabular? You could use a collection of vectors, but you will find that a structure called ***matrix*** is more useful. A matrix is essentially a rectangular collection of data. For instance, a matrix can store information that you typically see in a spreadsheet or table. A more abstract form of a matrix is an *array*, which Appendix D demonstrates.

9.6.2 Matrix Notation

Values stored inside matrices are commonly called ***matrix elements***. Row and column element positions of a matrix K are indicated by two indices i and j, respectively, as K_{ij}. For instance, matrix K has the following representation:

$$K = \begin{bmatrix} K_{11} & K_{12} \\ K_{21} & K_{22} \end{bmatrix} \tag{9.5}$$

This matrix organizes four elements into two rows and two columns. Element K_{11} corresponds to the information stored at row 1, column 1. Element K_{12} corresponds to the information stored at row 1, column 2, and so forth. *Square matrices*, like K, have equal numbers of rows and columns.

9.6.3 Matrix Terminology

Maple uses different terms for a mathematical matrix:

- *Matrix*: a matrix created from the **LinearAlgebra** package
- *matrix*: a matrix created from the **linalg** package

For this text, *matrix* refers to the mathematical notion of a matrix to avoid confusion. The term **Matrix** will match Maple's definition. I do not demonstrate **linalg**'s *matrix*.

9.6.4 Matrix Syntax

In Maple, there are many ways to enter a **Matrix**. I will focus on the form **Matrix([list1,list2,...,listn])**, which automatically creates a matrix by specifying each of a matrix's rows with lists:

- **list1** represents the first, or top, row.
- **listn** represents the last, or bottom, row.

The elements in each list represent the elements of the current row, as the following step demonstrates:

Step 241: Create a Square Matrix

> `K := Matrix([[2,-2],[-2,5]]);`   *Assign the matrix* $\begin{bmatrix} 2 & -2 \\ -2 & 5 \end{bmatrix}$ *to K.*

$$K := \begin{bmatrix} 2 & -2 \\ -2 & 5 \end{bmatrix}$$

Maple created a 4-by-4 matrix.

To see the options that Maple uses for a **Matrix** called *m*, enter **MatrixOptions(m)**. For more information, consult **?Matrix** and **?MatrixOptions**.

9.6.5   Matrix Palette and Shortcut Notation

If you cannot remember the syntax for creating a **Matrix**, use the Matrix Palette by selecting View → Palette → Matrix Palette, as shown in Figure 9.10. If you click any pattern, Maple will show a **Matrix** using shortcut notation:

- **<rowvec1, rowvec2, ..., rowvecn>** creates a **Matrix** from a collection of row vectors. Each row vector has the form **<elem1|elem2|...|elemn>**.
- **<colvec1|colvec2|...|colvecn>** creates a **Matrix** from a collection of column vectors. Each column vector has the form **<elem1, elem2, ..., elemn>**.

For instance, after typing **K:=** and selecting ▦ from the Matrix Palette, you would have the following input with placeholders:

Step 242: Create Matrix with Shortcut Notation

Enter **K:=**, *then use the Matrix Palette. Replace the placeholders with values:*
> `K := <<%? | %?> , <%? | %?>>;`

A highlighted placeholder indicates where the next input will go. You may use the Tab key to switch to a new placeholder. Press Enter after you have finished entering your values.

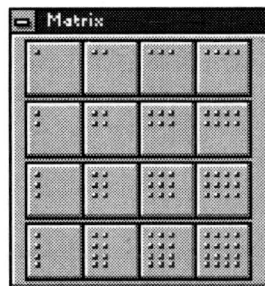

Figure 9.10   Matrix Palette

Maple allows you to enter shortcut notation without using the Matrix Palette. For instance, if you want to create a **Matrix** using row vectors, enter the following:

Step 243: Create Matrix with Rows

```
> mat := < <1|2> , <3|4> >;
```
*Create a **Matrix** using two row vectors.*

$$mat := \begin{bmatrix} 1 & 2 \\ 3 & 4 \end{bmatrix}$$

To create a **Matrix** using column vectors, do the following, instead:

Step 244: Create Matrix with Columns

```
> mat := < <1,2> | <3,4> >;
```
*Create a **Matrix** using two column vectors.*

$$mat := \begin{bmatrix} 1 & 3 \\ 2 & 4 \end{bmatrix}$$

Consult **?matpalette** and **?MVshortcut** for more information. To view a matrix, see How To...View...Structured Data Browser. You can even plot a matrix! Consult **?matrixplot**.

9.6.6   Matrix Extraction

To extract a matrix element from **row** and **col** of a matrix assigned to **name**, enter **name[row,col]**:

Step 245: Matrix Elements

```
> K := < <2|-2>, <-2|5> >:
```
Assign the matrix $\begin{bmatrix} 2 & -2 \\ -2 & 5 \end{bmatrix}$ *to K.*

```
> K[2,1], K[2,2];
```
$$-2, 5$$

Extract elements K_{21} and K_{22}.
Remember that matrix indices have the order row,col.

If you would like to extract a **Vector** or **Matrix** from another **Matrix**, use the syntax **name[L1,L2]**, according to the following rules:

- **L1** corresponds to rows.
- **L2** corresponds to columns.
- **L1** and **L2** can be integers, ranges, or lists.

In general, Maple will attempt to extract elements with indices that are common to both **L1** and **L2**. For instance, to exact the first row from *K* in Step 245, you would do the following:

Step 246: Matrix Elements

```
> K[1,1..2];
```
$$[2\ -2]$$

Extract the first row from K.
*Each row is also a **Vector**.*

As a trick for remembering the syntax, you might say, "For row 1, extract the elements from columns 1 to 2." If you use two ranges and/or lists, you will extract a smaller matrix, which is called a *submatrix*. For more information, see **?MVextract**, **?MVselect**, **?list**, and **?name**.

9.6.7 Matrix Assignment

Assuming you did not set your **Matrix** to option **readonly** (which means, *unchangeable*), you may reassign some, or all, of its elements. For instance, suppose you wish to replace a column with another inside a **Matrix**:

Step 247: Assign Vector Components

> `mat := Matrix([[1,2],[3,4]]):`   *Assign matrix $\begin{bmatrix} 1 & 2 \\ 3 & 4 \end{bmatrix}$ to mat.*

Replace the first column of mat with the column **Vector** $\begin{bmatrix} x \\ y \end{bmatrix}$:

> `mat[1..2,1] := Vector([x,y]);`

$$mat_{1,\,1..2} := \begin{bmatrix} x \\ y \end{bmatrix}$$ *Maple reports the assignment.*

> `mat;`   *Display the new contents of mat.*

$$\begin{bmatrix} x & 2 \\ y & 4 \end{bmatrix}$$ *Yes, you can include symbolic values in a* **Matrix**

For more information and rules, consult **?MVassignment**.

PRACTICE!

12. Assign the matrix $\begin{bmatrix} 1 & 2 & 3 \\ 4 & 5 & 6 \end{bmatrix}$ to A.

13. How many rows and columns does A have?

14. Extract elements A_{12} and A_{32}.

15. In A, replace the submatrix $\begin{bmatrix} 2 & 3 \\ 5 & 6 \end{bmatrix}$ with the matrix $\begin{bmatrix} 7 & 9 \\ 11 & 13 \end{bmatrix}$.

16. Replace the second row of matrix $\begin{bmatrix} 1 & 2 \\ 3 & 4 \end{bmatrix}$ with the vector $\begin{bmatrix} a & b \end{bmatrix}$.

9.7 ALGEBRA WITH VECTORS AND MATRICES

Linear algebra offers you powerful operations to perform on entire sets of data with vectors and matrices. The elegance derives from the fact that once you have stored the information, simple expressions with familiar operators can manipulate the multitude of values. This section provides primarily a summary of Maple's linear algebra operations and functions. Refer to the following help documents for more information:

- **?rtable_algebra**: operators for arrays, matrices, and vectors
- **?examples,LA_Syntax_Shortcuts**: examples of operations

- Mathematics…Linear Algebra…Linear Algebra Package…Standard…: common functions

This section demonstrates both the functions that Maple defines and the operator shortcuts, which you may use.

9.7.1 Vector Operations

Table 9.1 summarizes many vector operations that you will likely need. For instance, suppose that you need to add two vectors together. You may use the function **VectorAdd** or the add operator (**+**), which Maple defines as adding element-by-element:

Step 248: Vector Addition

```
> vector1 := Vector([1,2]):
```
Assign a vector.

```
> vector2 := Vector([-1,3]):
```
Assign another vector.

```
> vector1 + vector2;
```
*Add the two vectors. You could also use **VectorAdd**.*

$$\begin{bmatrix} 0 \\ 5 \end{bmatrix} \qquad \begin{bmatrix} 1 \\ 2 \end{bmatrix} + \begin{bmatrix} -1 \\ 3 \end{bmatrix} = \begin{bmatrix} 1 - 1 \\ 2 + 3 \end{bmatrix} = \begin{bmatrix} 0 \\ 5 \end{bmatrix}.$$

Consult the help documents for the functions in Table 9.1 for more information.

9.7.2 Matrix Operations

Table 9.2 summarizes many matrix operations that you will likely need. For instance, suppose you need to multiply a matrix by a vector. You may use the function **MatrixVectorMultiply**, which Maple defines differently for linear algebra:

TABLE 9.1 Vector Operations

Operation	Standard Math	Maple Notation
Equality	$V_i = U_i$	**Equal(A,B)**
Addition	$A_i + B_i = C_i$	**VectorAdd(A,B)** **A+B**
Scalar Multiplication	cA_i	**VectorScalarMultiply(A,c)** **c*A**
Dot Product	$A \cdot B = \sum_i A_i B_i$	**DotProduct(A,B)** **A . B**
Cross Product	$A \times B = \begin{bmatrix} A_2 B_3 - A_3 B_2 \\ A_3 B_1 - A_1 B_3 \\ A_1 B_2 - A_2 B_1 \end{bmatrix}$	**CrossProduct(A,B)**
Magnitude	$|A| = \sqrt{\sum_i A_i^2}$	**VectorNorm(A,2)**
Angle	$\theta = \cos^{-1} \dfrac{A \cdot B}{|A||B|}$	**VectorAngle(A,B)**

Step 249: Matrix-Vector Multiplication

```
> K := Matrix([[2,-2],[-2,5]]):
```
Assign a matrix $K = \begin{bmatrix} 2 & -2 \\ -2 & 5 \end{bmatrix}$.

```
> u := Vector([15,10]):
```
Assign a vector $u = \begin{bmatrix} 15 \\ 10 \end{bmatrix}$.

```
> MatrixVectorMultiply(K,u);
```
Find Ku.

$$[10, 20]$$

$$Ku = \begin{bmatrix} (2)(15) + (-2)(10) \\ (-2)(15) + (5)(10) \end{bmatrix} = \begin{bmatrix} 10 \\ 20 \end{bmatrix}$$

Order is crucial in linear algebra multiplication! For instance, when you need Mv, never compute vM! See **?LinearAlgebra[Multiply]** for a summary of matrix multiplication functions.

You eventually may tire of the long function names and wish to use a shortcut. Maple provides a linear algebra multiplication operator (**.**), but it requires care. Place at least one space before and after the dot. Otherwise, Maple will think you have constructed an expression of type float or range. For example, you could have entered **K . u** in Step 249 to perform the same operation. Consult **?dot** and the help documents for the functions in Table 9.2 for more information.

PRACTICE!

17. Evaluate $\begin{bmatrix} 1 & 2 \\ 2 & 1 \end{bmatrix} + 2\begin{bmatrix} -1 & 0 \\ 0 & -1 \end{bmatrix} - \begin{bmatrix} 3 & 1 \\ 1 & 3 \end{bmatrix}$.

18. Evaluate bAc where $A = \begin{bmatrix} 1 & 2 \\ 2 & 1 \end{bmatrix}$, $b = [3\ 1]$, and $c = \begin{bmatrix} -1 \\ 1 \end{bmatrix}$.

9.8   SOLVING SYSTEMS OF EQUATIONS

This section applies notions of linear algebra for solving a system of linear equations. Ensure that you have first entered **with(LinearAlgebra)** for this section.

9.8.1   Systems of Equations

You can convert a system of linear equations into *matrix form*, as shown in Figure 9.11. To do so, follow these steps:

*   Write each equation with the *unknowns* on the left and *knowns* on the right.
*   Write the equations on top of each other.
*   Identify the coefficients of the unknowns and place them in a matrix.
*   Place the unknowns and knowns in separate column vectors:

TABLE 9.2 Matrix Operations

Operation	Standard Math	Maple Notation
Equality	$A_{ij} = B_{ij}$	`Equal(A,B)`
Addition	$A_{ij} + B_{ij} = C_{ij}$	`MatrixAdd(A,B)`
		`A+B`
Scalar Multiplication	cA_{ij}	`MatrixScalarMultiply(A,c)`
		`c*A`
Vector Multiplication	$AX = \sum_j A_{ij}X_j$	`MatrixVectorMultiply(A,X)`
		`A . X`
Multiplication	$AB = \sum_k A_{ik}B_{kj}$	`MatrixMatrixMultiply(A,B)`
		`A . B`
Transpose	$A_{ij} = A_{ji}$	`Transpose(A)`
Inverse	A^{-1}	`MatrixInverse(A)`
Determinant	$|A|$	`Determinant(A)`
Identity Matrix	$I_{ij} = \begin{cases} 1 \text{ if } i = j \\ 0 \text{ if } i \neq j \end{cases}$	`IdentityMatrix(n)`

$$2u_1 - 2u_2 = 10$$
$$-2u_1 + 5u_2 = 20 \quad \Longrightarrow \quad \begin{bmatrix} 2 & -2 \\ -2 & 5 \end{bmatrix} \begin{bmatrix} u_1 \\ u_2 \end{bmatrix} = \begin{bmatrix} 10 \\ 20 \end{bmatrix}$$

Figure 9.11 Converting A System of Equations into Matrix Form

You may abbreviate the resulting form of the system as

$$Ku = p, \tag{9.6}$$

where matrix multiplication and equality are implied. Each term is described as follows:

System
$Ku = p$

The **system of equations** $Ku = p$ is a set of simultaneous linear equations.

Coefficient Matrix
$$K = \begin{bmatrix} 2 & -2 \\ -2 & 5 \end{bmatrix}$$

The **coefficient matrix** K collects the constants in front of unknowns:
- Square matrices, such as K, are used for the same number of unknowns and equations.
- Elements of these matrices typically reflect models' physical parameters.

Source Vector
$$p = \begin{bmatrix} 10 \\ 20 \end{bmatrix}$$

The **source vector** p applies modeled inputs, or "sources," to the system:
- In the spring example, these source terms are loads.
- Source values are typically known or assumed.

Solution Vector
$$u = \begin{bmatrix} u_1 \\ u_2 \end{bmatrix}$$

The **solution vector** u collects the unknown variables that you wish to find:
- The matrix formulation separates known and unknown variables.
- Manipulating the coefficient matrix and source vector with linear-algebra techniques, like Gaussian elimination, finds the unknowns.

9.8.2   Manual Solution

You can perform Gaussian elimination to solve the system of equations that Eq. (9.6) represents. But, why should you use the matrix formulation? Vectors and matrices store equation data in a compact form that computer programs can readily manipulate. Though more complex techniques exist, you can solve the matrix formulation with **row reduction**, which is a method that mimics Gaussian elimination, as demonstrated in Table 9.3.

The following steps illustrate row reduction:

- *Step* ①: Cast the equations into a matrix formulation.
- *Step* ②: Rewrite the system into a matrix that includes the source vector written to the right. You may draw a vertical bar to serve as a reminder to separate the coefficient matrix.
- *Step* ③: You can add a row to any other row, based on the rules of row reduction. So, add the top row to the bottom row. This process is equivalent to adding an equation to another equation within the given system.
- *Step* ④: You can multiply any row by any constant, based on the rules of row reduction. So, divide the top row by 2, and divide the bottom row by 3. You can perform this action in conjunction with adding rows to each other.
- *Step* ⑤: Keep performing row reduction until the coefficient matrix becomes the *identity matrix*, which is a matrix with values of 1 on the diagonal and 0 elsewhere. The final values in the right-hand column of the matrix represent the solution vector.

Note that a linearly dependent system of equations will yield at least one row that contains only values of zero.

9.8.3   Maple Solution

In Maple, to solve a linear system of equations $Ax = b$ for vector x, you should use **LinearSolve(A, b)**, given the following inputs:

- Matrix **A** is the coefficient matrix.
- Vector **b** is the source vector.

TABLE 9.3   Row Reduction

Step	Matrix Formulation	Operations	Results
①	$\begin{bmatrix} 2 & -2 \\ -2 & 5 \end{bmatrix}\begin{bmatrix} u_1 \\ u_2 \end{bmatrix} = \begin{bmatrix} 10 \\ 20 \end{bmatrix}$	$2u_1 - 2u_2 = 10$ $-2u_1 + 5u_2 = 20$	$2u_1 - 2u_2 = 10$ $-2u_1 + 5u_2 = 20$
②	$\begin{bmatrix} 2 & -2 & | & 10 \\ -2 & 5 & | & 20 \end{bmatrix}$		
③	$\begin{bmatrix} 2 & -2 & | & 10 \\ 0 & 3 & | & 30 \end{bmatrix}$	$\begin{aligned} 2u_1 - 2u_2 &= 10 \\ + \quad -2u_1 + 5u_2 &= 20 \\ \hline 0u_1 + 3u_2 &= 30 \end{aligned}$	$2u_1 - 2u_2 = 10$ $3u_2 = 30$
④	$\begin{bmatrix} 1 & -1 & | & 5 \\ 0 & 1 & | & 10 \end{bmatrix}$	$\frac{1}{2}(2u_1 - 2u_2 = 10) \rightarrow u_1 - u_2 = 5$ $\frac{1}{3}(0u_1 + 3u_2 = 30) \rightarrow u_2 = 10$	$u_1 - u_2 = 5$ $u_2 = 10$
⑤	$\begin{bmatrix} 1 & 0 & | & 15 \\ 0 & 1 & | & 10 \end{bmatrix}$	$\begin{aligned} u_1 - u_2 &= 5 \\ + \qquad u_2 &= 10 \\ \hline u_1 + 0u_2 &= 15 \end{aligned}$	$u_1 = 15$ $u_2 = 10$

For instance, solve the problem demonstrated in Figure 9.11 with the following steps:

Step 250: Solve Linear System with `LinearSolve`

> `K := Matrix([[2,-2],[-2,5]]):`   *Assign* $\begin{bmatrix} 2 & -2 \\ -2 & 5 \end{bmatrix}$ *to K.*

> `p := Vector([10,20]):`   *Assign* $\begin{bmatrix} 10 \\ 20 \end{bmatrix}$ *to p.*

> `u := LinearSolve(K,p);`   *Solve* $Ku = p$ *for u.*

$$u := \begin{bmatrix} 15 \\ 10 \end{bmatrix}$$ *Maple found* $u = \begin{bmatrix} 15 \\ 10 \end{bmatrix}.$

For more information about a variety of options you can supply, see **?LinearSolve**.

9.8.4   Linearly Dependent Systems

Generally, you should ensure that the number of equations equals the number of unknowns. Thus, be aware of duplicated equations that create a linearly dependent system. Given too few equations, or a linearly dependent system, **linsolve** produces parametric solutions in terms of "_t" variables. For example, can you solve the system composed of just $x_1 + x_2 = 2$ and $2x_1 + 2x_2 = 4$? Note how the second equation is a multiple of the first by a factor of 2:

Step 251: Solve Linearly Dependent System

> `A := Matrix([[1,1],[2,2]]):`   *Assign* $\begin{bmatrix} 1 & 1 \\ 2 & 2 \end{bmatrix}$ *to A.*

> `b := Vector([2,4]):`   *Assign* $\begin{bmatrix} 2 \\ 4 \end{bmatrix}$ *to b.*

> `x := LinearSolve(A,b);`   *Solve* $Ax = b$ *for x.*

$$x := \begin{bmatrix} 2 - \_t0_2 \\ \_t0_2 \end{bmatrix}$$ *Maple writes the solution in terms of parameters.*

You might see Maple write the solution in terms of another parameter, as well. In either case, to eliminate linear dependency, replace one of the equations, or give one of the parameters a value.

9.8.5   Verification

Check your work! After finding the solution vector u from $Ku = p$, multiply K by u to confirm that you generate p. To directly compare Ku with p, use **Equal(M1,M2)**, where **M1** and **M2** are two vectors or two matrices:

Step 252: Check Linear System of Equations

Use **K**, **u**, *and* **p** *from Step 250.*

```
> Equal(Multiply(K,u),p):
                    true
```
Multiply matrix K by vector u and check if Ku = p.
Yes, Maple found the right solution.

See **?LinearAlgebra[Multiply]** and **?LinearAlgebra[Equal]** for more information.

PRACTICE!

19. Solve the system of equations $\{x + y = 2, x - y = 0\}$ with **solve**.
20. Now, try a graphical approach.
21. Convert the system to matrix form, and use **LinearSolve** to solve it.
22. Finally, check your solutions by back substituting the solution vector.

PROFESSIONAL SUCCESS: GO TO CLASS!

Do you wish to perform well on homework, projects, and tests? Then, it's best to attend class. Here are some reasons and related tips:

- Record announcements: Where are you going to find out the assignment? Go to class. What material does the homework cover? Go to class. What material will be tested? Go to class. Students often ask me when homework is due on the very date that it is due!

- Learn beforehand: Some classroom boredom arises from bafflement and wandering thoughts. Prepare ahead of time! Skim the textbook a few times. After all, repetition aids learning.

- Listen carefully: Listen to your professor in class. Guess what usually shows up on tests? Students who stay alert and listen usually perform well.

- Take notes: Combine careful note taking with good listening. Let your professor's voice guide your writing. Later, combine assigned reading with lecture notes.

- Ask questions: When material confuses you, don't be afraid to raise your hand! Half your class probably will thank you. Someone is paying good money for your education, so make the most of it.

- Now, go to class!

KEY TERMS

coefficient Matrix	constructor	engineering
gaussian elimination	last name evaluation	linear algebra
linear system	linearly dependent	linearly independent
matrix	matrix elements	module
nonlinear system	row reduction	rtable
solution vector	source vector	system
system of equations	vector	vector components

SUMMARY

- You can build a model from a set of simultaneous equations.
- Linear equations do not have unknown variables with a power greater than one.

- Linear independency means that a set of simultaneous equations is solvable for the unknown variables if every equation is linear and not a duplicate of another.
- Linear dependency means that a set of simultaneous equations is not solvable for the unknown variables if at least one equation is a duplicate of another.
- You can solve a set of linear and nonlinear set of equations with **solve**.
- You can solve a set of linear and nonlinear set of equations by finding the intersecting point(s) on a plot.
- Linear algebra is a form of mathematics that deals with multiple values, which helps you to solve a system of equations.
- Maple has two main packages for linear algebra, which are **LinearAlgebra** and **linalg** and must be accessed with the **with** function.
- You should use **LinearAlgebra** instead of **linalg**.
- The **LinearAlgebra** package represents vectors and matrices as rectangular tables.
- You may represent columns and rows of expressions with vectors.
- You may represent tables of expressions with matrices.
- You may perform operations on vectors and matrices.
- To set up a system of equations, you write the coefficients of the unknowns in matrix, the unknowns in a column vector, and the knowns in another column vector.
- To solve a system of equations, use **LinearSolve(*CoefficientMatrix, KnownVector*)**.
- You should check your results by multiplying the solution vector with the coefficient matrix and comparing the result with the source vector.

Problems

1. Distinguish a scalar quantity from a vector quantity.
2. Describe the difference between Maple's lists and vectors.
3. Explain the difference between vectors and matrices.
4. Considering matrix size, when can you multiply two matrices?
5. Assign the following vectors:

$$x = \begin{bmatrix} 3 \\ -2 \\ 0 \end{bmatrix}$$

and

$$y = \begin{bmatrix} -5 \\ 2 \\ 4 \end{bmatrix}.$$

Evaluate the following expressions:

(a) $x + y$
(b) $x - y$

(c) $x \cdot y$

(d) $x \times y$

(e) $(2x + 3y) \cdot (-x)$

(f) $|x|$

(g) $|x \times y|$

(h) Find the angle between x and y.

Check your work. Ensure that Maple prints the solutions.

6.   Assign the vector

$$c = \begin{bmatrix} 1 \\ 2 \end{bmatrix}$$

and matrices

$$A = \begin{bmatrix} 1 & 2 \\ 2 & 1 \end{bmatrix}$$

and

$$B = \begin{bmatrix} -2 & -1 \\ 3 & 0 \end{bmatrix}.$$

Evaluate the following expressions:

(a) $A + B$

(b) $A - B$

(c) AB and BA. Do the results differ? Why or why not?

(d) BB^{-1}. Do you obtain an identity matrix? Why or why not?

(e) $|A|$

(f) B^T

(g) Ac. Can you also evaluate cA? Why or why not?

(h) ABA. Hint: Beware of the order of operations.

Check your work. Ensure that Maple prints the solutions.

7.   Is the system of equations $\{x + 2y = 5, -2x - 4y = -10\}$ linearly dependent or independent? Why or why not? Check for linear independence or dependence using Maple.

8.   Can you solve the system of equations $\{x - y = 2, \sin x + y = 10\}$ using linear solution techniques? Why or why not? Solve for x and y and verify your results.

9.   Use any method you see fit to solve the following systems of equations. Be sure to verify your results:

(a) $\begin{Bmatrix} x + 2y = 4 \\ -2x + y = -3 \end{Bmatrix}$

(b) $\left\{\begin{array}{l} x + 2y = 13 \\ -2x - 4y = -26 \end{array}\right\}$

(c) $\left\{\begin{array}{l} 10x - 17y = 13.2 \\ -6.5x + 0.11y = -7.1 \end{array}\right\}$

(d) $\left\{\begin{array}{l} x + y + z = 6 \\ x - y - z = -4 \\ -x - y + z = 0 \end{array}\right\}$

(e) $\left\{\begin{array}{l} x + 2y - 3z = 14 \\ -3y + 11z = -2 \\ x = 0 \end{array}\right\}$

(f) $\left\{\begin{array}{l} w + \ -2x - 1y + 3z = 4 \\ -x + 4y + 5z = 10 \\ w + x + 7z = -9 \\ -4w - x - 3y + 12z = -8 \end{array}\right\}$

10. Refer to Problem 18 in Chapter 4. Consider the equation

$$ Z = R - \frac{j}{\omega C}. \qquad (9.7)$$

The components of the complex number Z form a vector called a *phasor* when plotted along a real and imaginary axis. Treat the real axis as horizontal, and the imaginary axis as vertical. Answer the following problems using the information in Problem 18 in Chapter 4:

(a) Identify and assign the horizontal and vertical components of Z.

(b) Find the magnitude of the vector Z.

(c) Find the angle at which Z rotates from the real, or horizontal, axis.

11. You can use vectors to solve *equilibrium* problems, which involve static bodies that do not move. The two cables shown in Figure P9.12 balance a suspended weight (W) of 100 N:

Do the following tasks:

(a) Resolve both cable forces into horizontal and vertical components. Use trigonometry and the indicated axes in Figure P9.12. For instance, $(T_1)_x = -T_1 \cos 30°$. Report your results as text in an execution group.

(b) Assign the variables T_{1x} and T_{1y} to the horizontal and vertical components of cable force $\vec{T}_1$, respectively. If you prefer, you may skip including the subscripts.

(c) Assign the variables T_{2x} and T_{2y} to the horizontal and vertical components of cable force $\vec{T}_2$, respectively. If you prefer, you may skip including the subscripts.

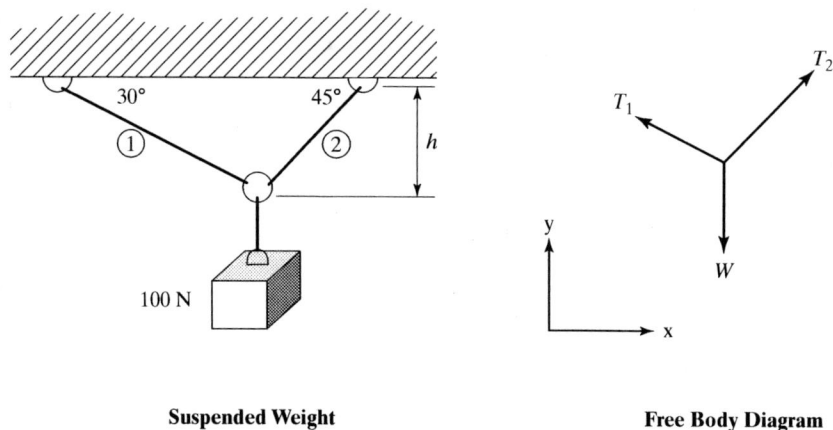

Suspended Weight   **Free Body Diagram**

Figure P9.12   Equilibrium Problem

(d) Report the value of W_x, the horizontal component of force W. Report the value of W_y, the horizontal component of force W. Assign the appropriate values to the variables W_x and W_y.

(e) The sum of the forces in the x direction must balance. Express this relationship in Maple using horizontal components of each force. Hint: Assign the equation $T_{1x} + T_{2x} + W_x = 0$ to the variable *Eqn1*.

(f) The sum of the forces in the y direction must balance. Express this relationship in Maple using vertical components of each force. Hint: Assign the equation $T_{1y} + T_{2y} + W_y = 0$ to the variable *Eqn2*.

(g) Solve for both cable forces. Check your work by comparing the results from another approach that you should determine. Do you achieve the same answers?

(h) Project: Let the height $h = 1$ m in Figure P9.12. Assume that you cannot allow each cable's displacement to exceed 1% of h (1 cm). Assuming circular cross-sections for the cables, find the maximum diameter of each cable using Maple.

Hints:

• Assume that cables behave in accordance with Hooke's Lab for axial members,

$$T = \frac{AE}{L}u, \tag{9.8}$$

where force T stretches the cable with displacement u. The other parameters are circular cross-sectional area A, length L, and Young's modulus E (force/area).

• Assume that both cables have the same value of $E = 300 \times 10^6$ Pa (N/m^2). Beware that each cable might have different parameters. Use different names for areas, like *A1* and *A2*, and so forth. You will have to solve for the lengths and iterate for displacements and areas.

- Assume that the weight moves with a "small" displacement Δ. Therefore, the deformed geometry does not appreciably alter the original geometry. Better yet, you can apply original equilibrium forces to the deformed system.

- Both cables must have compatible displacements. That is, the cables must stay connected. You should resolve displacement Δ into x and y components by letting $u = \Delta_x$ and $v = \Delta_y$. Sum the contributions for each cable as $\Delta_1 = u_1 + v_1$ and $\Delta_2 = u_2 + v_2$, according to the geometry.

10

Introduction to Calculus

10.1 INTRODUCTION

Models represent mathematical abstractions of physical systems. Analyzing the equations in models helps engineers and scientists predict, analyze, and design the real physical systems. This section introduces how calculus helps model changes in physical systems.

OBJECTIVES

After reading this chapter, you should be able to

- Understand how calculus models change in physical states
- Identify smoothness and continuity on functions
- Understand the notion of a limit
- Demonstrate how derivatives model changes by finding slopes
- Demonstrate how integrals model sums by finding areas and volumes
- Demonstrate how differentiation and integration are inverse operations

10.1.1 Change

Models encompass many features of physical systems, like *change* in space, time, and other physical characteristics. Consider a rocket shooting towards space. Engineers must account for increasing velocity, decreasing gravity, decreasing mass of fuel, vibrating components, and many other factors. A form of mathematics called ***calculus*** provides a methodology for modeling these changes in equations.

10.1.2 Maple and Calculus

Maple's standard calculus functions are described in Mathematics...Calculus.... This chapter will use many of those routines along with Maple's new Student package, which is described in **?Student** and **?Student[Calculus1]**. See Student Package...Calculus 1...Overview for the full listing of commands. You will also find excellent calculus examples in **?examples[index]**. Note that **Student** supersedes the older **student** library package, which also provides commands for learning calculus. To access the functions, enter **with(Student[Calculus1])** for this chapter.

Step 253: Student Library Package

```
[> restart:                              Restart your Maple session.

[> with(Student[Calculus1]):      Access functions in the student library package.
```

10.1.3 Related Packages

Because Maple is extremely powerful, there is way too much material to cover in this text! Two packages that you will find useful as you progress through your engineering studies are

- **?VectorCalculus**, Mathematics...Vector Calculus...
- **?VariationalCalulus**, Mathematics...Calculus of Variations...

Other packages are mentioned throughout this chapter.

PROFESSIONAL SUCCESS: TAKE BABY STEPS...

Great feats rarely occur overnight, and that goes for trying to learn Maple in its entirety in one evening. You should break projects into "baby steps," or small steps whose sum encompasses your goals. Consider these aspects of this process:

- Conception: Imagine your goals. As you would when writing a paper or project report, brainstorm your ideas. Eventually, your objectives might shift, but you need a starting point. Starting is half the battle!
- Clarification: Stay realistic. Follow guidelines and understand your main objectives. Double your time expectations when struggling to meet deadlines.
- Baby Steps: Attempt everything all at once, and you likely will never finish. Instead, outline your

tasks. Decompose each step into smaller steps. After all, smaller tasks are easier to complete. In time, doing each step, however small, will add up to finishing the larger task.

- Procrastination: Start with the easiest tasks. Whether your first job involves cleaning your desk or labeling your files, simply starting will provide necessary momentum for more important tasks.
- Work: Set a schedule and stick to it. When stuck, never be afraid or ashamed to seek help. Everyone needs help sometimes!
- Life: Balance your life. Reward yourself when finishing tasks. Take a walk. Call a loved one. People need companionship and personal time. Stay human, and your goals will flourish.

10.2   FUNCTIONS

Functions embrace change by modeling different physical states given different variable values. Consider Hooke's Law, $p = ku$, which models the interaction between the displacement u and the force p in a linear spring. Changing u or p produces new physical states, which you can plot as points in (u, p) coordinates. This section studies how functions vary between these states.

10.2.1   Limits

Suppose that you to investigate the function behavior of $f(x)$ near a point $x = c$. As $x \rightarrow c$ (that is, "as x approaches a value c"), $f(x)$ approaches a value L. The specific value of $f(x)$ at L is called the **limit**, which can be expressed as

$$\lim_{x \to c} f(x) = L. \tag{10.1}$$

Why the word *limit*? The limit represents the biggest or smallest value the function could possibly have at a point, even if the function cannot be evaluated at that point.

Consider the function $\ln(x)$. You cannot actually find the natural logarithm of a nonpositive number. However, a plot of $\ln(x)$ shows that the value approaches a limit of negative infinity as x approaches 0 from the right. In Maple, to find the limit of **f** as **x** approaches **c**, use **limit(f,x=c)**:

Step 254: Limits

> **limit(ln(x),x=0);**   *Find* $\lim\limits_{x \to \frac{x}{2}} \ln x.$

                              $-\infty$ *To confirm, see* **plot(ln(x),x)**.

If you want to specify the direction from which to take the limit, use **limit(f,x=c,dir)**, where **dir** is **left**, **right**, **real**, or **complex**. For more information, consult **?limit**, and Mathematics...Calculus...Limits.... From the **Student** package, you should see **?LimitRules** for functions that give you a step-by-step explanation.

10.2.2   Continuity

Continuous functions have limits that match function values:

$$\lim_{x \to c} f(x) = f(c). \tag{10.2}$$

In Figure 10.1, $f(x)$ is continuous between points A, B, and C. However, at points C and C' the function $f(x)$ has no unique value, and thus, loses continuity. Moreover, the limit of $f(x)$ at points C and C' depends on whether you approach those points from the left or the right. Except for points C and C', however, all other points along $f(x)$ are continuous. Consult **?discont**, **?fdiscont**, **?iscont**, and **?singular** for more information.

10.2.3   Smoothness

Smooth functions contain no abrupt changes in slope. As shown in Figure 10.1, the slopes of $f(x)$ to the left and right of point B change, but point B suffers a slopeless fate. Point B forms a cusp, which is a sharp point with no slope. In general, smooth functions lack cusps. Functions, like $f(x) = mx + b$ and $f(x) = \sin x$, have continuous and defined slopes, and thus, are good examples of smooth functions. Functions, like $f(x) = |x|$, have undefined slopes at some locations, and are thus nonsmooth.

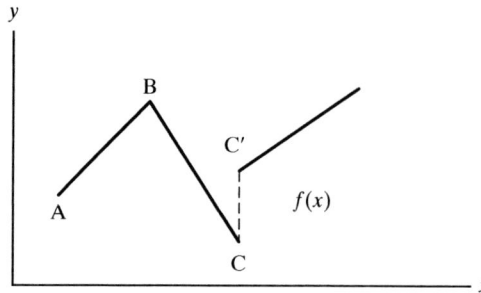

Figure 10.1 Non-Continuous Function

10.2.4 Piecewise Functions

You might wish to investigate **?piecewise** to create *piecewise* functions of your own that resemble Figure 10.1. A piecewise function is composed of a set of functions that depend on different ranges of the independent variable. By combining functions with different slopes, you can create points that lack smoothness. Furthermore, by specifying disconnected ranges for the independent variable, you will create a noncontinuous function as well.

PRACTICE!

1. Plot tan(x) for $0 \leq x \leq 2\pi$. Hint: See **?plot** regarding the **discont** option.
2. Find

$$\lim_{x \to \frac{\pi}{2}} \tan x$$

   Hint: See **?limit** and **?limit[dir]**.
3. What values of x make tan(x) discontinuous? Hint: See **?discont**.

10.3 SLOPE

Derivatives measure rate, or how quickly functions change. Before covering derivatives, this section first introduces the related and familiar concept of slope, which also measures change.

10.3.1 Change and Slope

Slope is a measure of change in elevation of a surface or line. You might have encountered various definitions of slope for a linear equation $y = f(x)$:

$$slope = \frac{rise}{run} = \frac{\Delta y}{\Delta x} = \frac{y_2 - y_1}{x_2 - x_1}. \tag{10.3}$$

Many formulas employ slope to help model physical systems that change. For instance, consider the plot and formulas of Hooke's law, as shown in Figure 10.2. The spring stiffness k measures how the load p changes linearly with the displacement u.

10.3.2 Tangents

Linear equations in the form $y = mx + b$ have a constant slope m. Varying x and y does not change m. What is the slope of a nonlinear equation? For nonlinear equations, you can measure slope at individual points by drawing a tangent to the curve. A tangent's slope matches the original equation's slope at the point where the tangent touches the original equation.

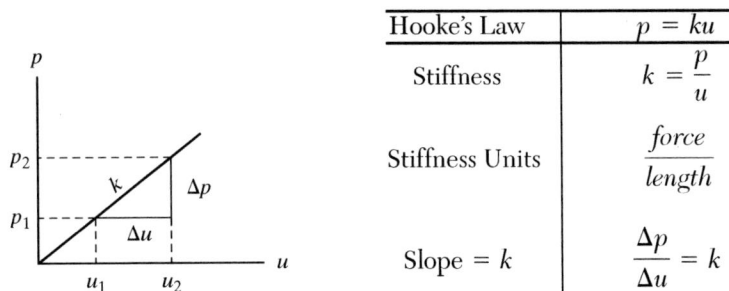

Hooke's Law	$p = ku$
Stiffness	$k = \dfrac{p}{u}$
Stiffness Units	$\dfrac{force}{length}$
Slope $= k$	$\dfrac{\Delta p}{\Delta u} = k$

Figure 10.2   Slope for Hooke's Law

The spring example can demonstrate a nonlinear curve and tangent lines. The equation

$$E_s = \frac{1}{2}ku^2,\tag{10.4}$$

which models the potential energy of a spring, is nonlinear because of the u^2 expression. Assuming a value of $k = 2$, suppose that you wish to draw a tangent to the curve at $u = 4$. Use Maple's **Tangent** function, which is quite versatile. I will demonstrate the syntax **Tangent(*expr*,*var=c*,*range*,*opts*)**:

- **expr** must be an expression or function in terms of **var**.
- **var=c** is the point on the curve on which you wish to draw the tangent.
- **range** is the range of values of the horizontal axis expressed as **low..high**.
- **opts** is a sequence of options that change the form of output, appearance, and title.

The following input draws the tangent to $f(x) = x^2$ at $x = 4$:

Step 255: Show Tangent

> **Tangent(u^2, u=4, 0..6, output=plot);**   *Find the tangent of* f(u) = u^2
at u = 4.

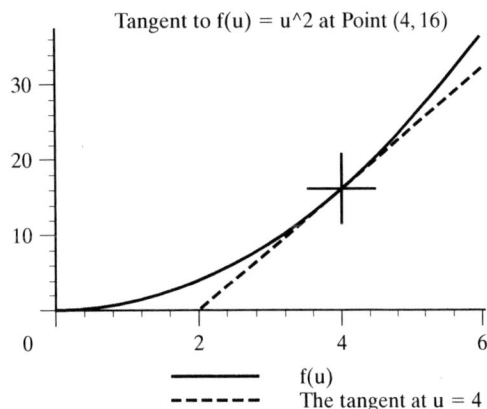

Actually, I modified my plot to make it look better for the book.

*I changed pretty much all of the options that **Tangent** has:* **output, functionoptions, pointoptions, tangentoptions,** *and* **title**.

Curious how I did this? See **?Calculus1[Tangent]** *and* **?Calculus1[plot_options]** *to unravel the mystery.*

At $u = 4$, you can determine the slope of $E_s = 8$, as shown in Step 255:

$$slope = \frac{f(4^2) - f(0^2)}{4 - 2} = \frac{16}{2} = 8. \tag{10.5}$$

You can have Maple figure that value, too!

Step 256: Find Slope

```
> Tangent(x^2, x=4, output=slope);
                    8
```

For more information on **Tangent** and its variety of options, see **?Calculus1 [Tangent]**.

PRACTICE!

4. Show the commands, including all options, for the plot in Step 255.

5. Draw a tangent line at $x = \dfrac{\pi}{2}$ to the function $f(x) = \sin(x)$.

6. Determine the equation for slope of the quadratic equation $y = ax^2 + bx + c$ at $x = 0$.

10.4   DERIVATIVES

This section introduces differentiation, or how derivatives produce a function's slope.

10.4.1 Relating Slopes and Derivatives

Calculus provides an analytical method for finding tangents and, in turn, slopes. Consider the curve $f(u)$, as shown in Figure 10.3. To find the slope of line ab, use Eq. (10.3):

$$m_{ab} = \frac{f(u_b) - f(u_a)}{u_b - u_a} = \frac{f(u_a + \Delta u) - f(u_a)}{(u_a + \Delta u) - (u_a)} = \frac{f(u_a + \Delta u) - f(u_a)}{\Delta u}. \tag{10.6}$$

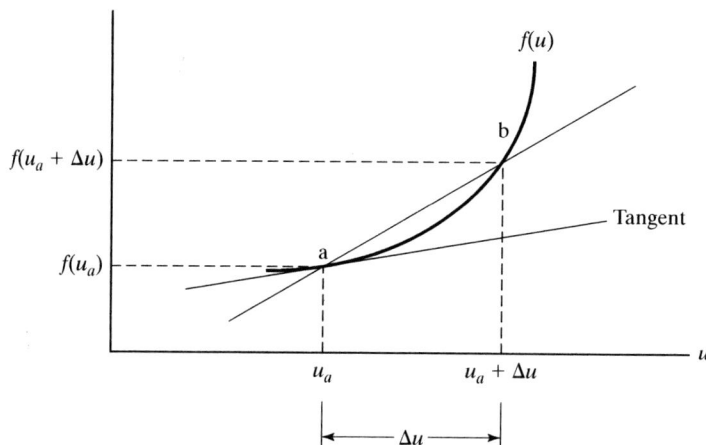

Figure 10.3   Finding Slopes

Next, apply mathematical limits such that Δu shrinks as $\Delta u \to 0$. Then, the line *ab* approaches the tangent to point *a*:

$$m_{\text{Tangent}} = \lim_{\Delta u \to 0} \frac{f(u_a + \Delta u) - f(u_a)}{\Delta u} \tag{10.7}$$

Equation (10.7) represents the formula for a **derivative**, which is a mathematical measure of slope, or the instantaneous rate of change of a function at a point. Note that derivatives exist only where a function is smooth and continuous.

Maple's **Calculus1** package can help you to explore the meaning of Eq. (10.7) with the **NewtonQuotient** function, which has a similar syntax as **Tangent**:

Step 257: Find the Slope at a Point

```
> opts := output=animation, 'h'=1, iterations = 10:    Set some options.

> NewtonQuotient (u^2,u=4,0..6,opts);                   You can play this animation!
```

Newton Quotient of u^2 at u = 4

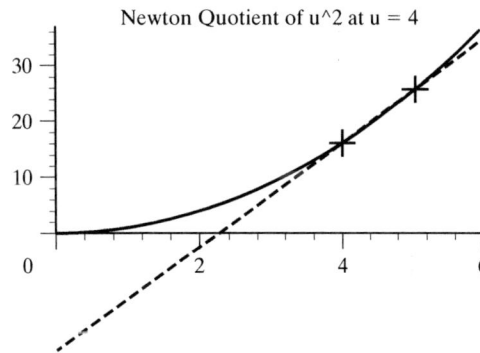

Actually, I modified my plot to make it look better for the book.

I changed many of **NewtonQuotient** *'s options:* **output**, **'h'**, **iterations**, **pointoptions**, **functionoptions**, **derivativeoptions**, **quotientoptions**, *and* **title**.

Curious how I did this? See **?Calculus1[NewtonQuotient]** *and* **?Calculus1[plot_options]**.

As you "play" the animation in Step 257, the two crosses will converge as Maple iterates toward the slope at the point $u = 4$. Consult **?Calculus1[NewtonQuotient]** for more information.

Using $\Delta y/\Delta x$ only approximates slope, but you can improve your accuracy. Let $y = f(x)$. The **first derivative** of y with respect to x finds the exact, *instantaneous* slope anywhere on $f(x)$:

$$\textbf{First Derivative}: y' = f'(x) = \frac{dy}{dx} = \frac{d}{dx}f(x) \tag{10.8}$$

Whereas Δ indicates a large change, d indicates an infinitesimal change in a variable. Thus, derivatives provide exact slopes.

10.4.2 Physical Interpretation

Derivatives determine rates, or how functions change, and thus, provide useful tools for analysis and design. Consider the formula for the potential energy of a spring:

$$E_s = \frac{1}{2}ku^2. \tag{10.9}$$

The equation $E_s = f(u)$ models a physical system, where some parameters, like E_s and u, vary. Other parameters, like k, are thought of as *constants*, and thus do not vary for a particular model. Taking the derivative E_s' measures how E_s changes when the displacement u changes. For example, the slope of E_s is $E_s' = 2u$ for $k = 2$. The derivative $2u$ now provides a new formula for measuring how E_s changes as u varies.

10.4.3 Maple Syntax

To find the derivative of **func** with respect to **var**, enter **diff(func, var)**. For instance, given a function $y = f(x) = ax^2$, you would enter **diff(y,x)** or **diff(f(x),x)** to compute y'. Try the spring example using Eq. (10.4):

Step 258: First Derivative—Cleanup

```
> k:='k': u:='u':
```
                                                        Clear k and u assignments.

Step 259: First Derivative—Assign

```
> Es := (1/2)*k*u^2:
```
                                                        Assign $\frac{1}{2}ku^2$ to E_s.

Step 260: First Derivative—Evaluate

```
> diff(Es,u);
```
                                                        Find the first derivative E_s' of $\frac{1}{2}ku^2$.

$$ku$$
                            This expression is the slope of $\frac{1}{2}ku^2$ at any point u.

In functional notation, use parentheses such that $E_s(u) = \left(\frac{1}{2}\right)ku^2$:

Step 261: First Derivative—Functional Notation

```
> Es := u->(1/2)*k*u^2:
```
                                                        Assign $\frac{1}{2}ku^2$ to $E_s(u)$.

Step 262: First Derivative—Evaluate

```
> diff(Es(u),u);
```
                                                        Find the first derivative $E_s'(u)$.

$$ku$$
                            This expression is the slope of $\frac{1}{2}ku^2$ at any point u.

Also, try **implicitdiff(eqn, y, x)** to directly differentiate the equation $f(x, y) = c$, where both y and x vary independently, given constant c. For additional help, see **?Calculus1[DiffRules]** and **?Calculus1[DerivativePlot]**.

10.4.4 Higher-Order Derivatives

Differentiating a derivative creates a **higher order derivative**. For instance, a second derivative measures how much a first derivative changes with respect to the independent variable. Thus, the second derivative of y with respect to x yields

$$y'' = \frac{d}{dx}\left(\frac{dy}{dx}\right) = \frac{d^2y}{dx^2} \tag{10.10}$$

To find a second derivative, enter **diff(*func*, *var*, *var*)** or just **diff(*func*, *var*$2)**. The input **var$2** abbreviates the sequence **var, var**:

Step 263: Second Derivative

```
> diff(Es(u),u,u);
```
k

Evaluate $E_s''(u)$.
Also, try **diff(diff(Es(u),u),u)**.

To find an nth-order derivative, enter **diff(*func*, *var*$n)**. Table 10.1 organizes common notations for many types of derivatives. Beware of the syntax change when using functional notation.

10.4.5 Partial Derivatives

When functions have more than one independent variable, Maple takes **partial derivatives**, which take the form $\frac{\partial^{(n)}}{\partial x_1 \partial x_2 \ldots \partial x_n} f(x)$. However, when you have only one independent variable, you can assume the following relationship:

$$\frac{\partial}{\partial x} f(x) = \frac{d}{dx} f(x) \tag{10.11}$$

For example, assume that some function f has two indepent variables x and y:

TABLE 10.1 Differentiation

Operation	Standard Math	Maple Notation
First derivative of y with respect to x	$\frac{dy}{dx}$ and y',	`diff(y,x)`
First derivative of $f(x)$ with respect to x	$\frac{d}{dx}f(x)$ and $f'(x)$	`diff(f(x),x)`
First derivative of y with respect to time, t	$\frac{dy}{dt}$ and $\dot{y}$	`diff(y,t)`
Second derivative of y with respect to x	$\frac{d^2y}{dx^2}$ and y''	`diff(y,x,x)`
		`diff(y,x$2)`
Second derivative of y with respect to t	$\frac{d^2y}{dt^2}$ and $\ddot{y}$	`diff(y,t,t)`
		`diff(y,t$2)`
nth derivative of y with respect to x	$\frac{d^ny}{dx^n}$ and $y^{(n)}$	`diff(y,x,x,...)`
		`diff(y,x$n)`
Partial derivative of $f(x, y)$ with respect to x	$\frac{\partial}{\partial x}f(x,y)$	`diff(f(x,y),x)`
Partial derivative of $f(x, y)$ with respect to y	$\frac{\partial}{\partial x}f(x,y)$	`diff(f(x,y),y)`
Partial derivative of $f(x, y)$ with respect to x and y	$\frac{\partial^2}{\partial x \partial y}f(x,y)$	`diff(f(x,y),x,y)`
		`diff(f(x,y),y,x)`

Step 264: Partial Derivatives:

> `diff(f(x,y),x);`   *Find the derivative of* `f(x,y)` *with respect to* x.

$$\frac{\partial}{\partial x}f(x,y)$$ *Maple does not evaluate inert functions.*

See **?diff** for more information.

10.4.6 Miscellaneous

Consult these references for more information:

- Differential Operators: **?D**, **?operators[D]**, **?difforms**
- Vector Calculus: **?curl**, **?diverge**, **?grad**, **?laplacian**
- Differential Equations: Mathematics…Differential Equations…, **?dsolve**

PRACTICE!

7. Find the first derivative of $f(x) = ax^2 + bx + c$ with respect to x.
8. What is the slope of $f(x) = ax^2 + bx + c$ at $x = 2$? Do not use **Tangent**.
9. Find the second derivative of $f(x) = ax^2 + bx + c$ with respect to x.
10. What is the first derivative of $f(x, y) = xy^2 + x^2y$ with respect to x?

10.5   AREA

The concepts of area that are discussed in this section will clarify the integration methods in the next section.

10.5.1 Summation

Let s represent a sequence of values:

$$s = s_1, s_2, \ldots, s_n. \tag{10.12}$$

The summation of all values s is represented using sigma notation Σ as

$$\sum_{i=1}^{n} s_i = s_1 + s_2 + \ldots + s_n, \tag{10.13}$$

where s_i denotes each individual element of the sequence s. The index i ranges from 1 to n, where n is the last element in the sequence. To calculate the summation of a sequence **seq** over a particular range **range**, enter **sum(seq, index=range)**:

Step 265: Summation

> `Seq := 2,4,6,8:`   *Assign a sequence to* Seq.

> `i := 'i':`   *Clear any assignment on* i.

> `sum(S[i],i=1..4);`   *Sum each value in the sequence* s_1, s_2, s_3, s_4.

$$20$$ $2+4+6+8 = 20.$

If Maple reports an error, go back and unassign the index **i** with **i:='i'** before evalu-ating the summation or enter **sum(S['i'],'i'=1..4)**.

10.5.2 Approximate Area

Suppose that you wish to approximate the ***area***, or the amount of space, beneath the function $p = ku$ down to the u-axis and between two values of u. For instance, assume that $k = 2$ and $0 \le u \le 4$. To approximate the area below ***expr*** with boxes in a given ***range***, use **ApproximateInt(*expr*, *range*, *options*)**. The "Int" refers to *integration*, which I describe later. In the example that follows, I have chosen one op-tion, **output=plot**, to graphically demonstrate the midpoint, or "rectangle," method of approximating area:

Step 266: Approximate Area with Rectangles

Set plot options. I'm only showing the major options that affect the appearance. Some of these options actually are default values:

```
> opts := labels=[",""], legend=" ", method=midpoint,
         outline=false, partition=6, refinement=halve,
         showarea=false, showfunction=true,
         showpoints=false, title="Approximate Area":
```

```
> ApproximateInt(2*u,u=0..4,opts,output=plot,opts);
```

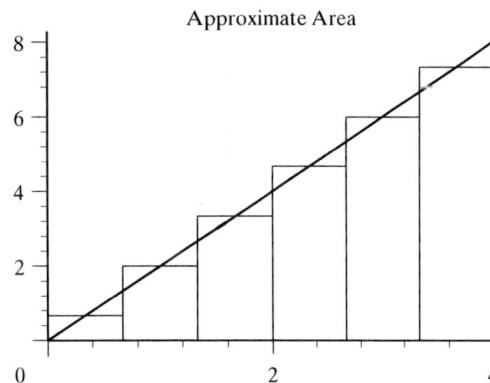

Approximate Area

Actually, I modified my plot to make it look better for the book.

Curious how I did this? This function has many options you can modify.

See **?Calculus1[Approximate Area]** *and* **?Calculus1[plot_options]** *to unravel the mystery.*

To evaluate the approximate area in the defined region as summation, use the option **output=sum**. Otherwise, for a numerical answer, use the option **output=value**, which also is the default:

Step 267: Calculate Approximate Area

Approximate the area as a sum of rectangles:

```
> ApproximateInt(2*u, u=0..4, partition=6, output=sum);
```

$$\frac{2}{3}\left(\sum_{i=0}^{5}\left(\frac{4}{3}i + \frac{2}{3}\right)\right)$$

This sum is the total area of all six rectangles.

> *Determine the numerical value for the area:*
> ```
> > ApproximateInt(2*u, u=0..4, partition=6, output=value);
> ```
>   16   *Using middle boxes generates exact results for straight lines.*

Because $2u$ is a straight line, the "approximation" is actually perfect. The Practice problems that follow demonstrate that the approximate and correct values for curves generally are different. Consult **?ApproximateInt** for more information, especially on more methods that improve the approximation. See **?value** and **?Sum** for related functions.

PRACTICE!

11. Estimate the area below $f(x) = x^2$ between $0 \le x \le 4$ with a plot and a numerical value.

12. You will learn that **int(x^2,x=0..4)** will find the exact area. Enter this input and compare the result with your approximation.

13. If the results differ, how could you improve the approximation?

10.6   INTEGRALS

Finding area is just one application of *integration*, which is discussed in this section.

10.6.1   Relating Area and Integrals

Consider the general function $f(u)$ shown in Figure 10.4. You may approximate the area below $f(u)$ with a sequence of rectangular slices, which each have an area of

$$A_i = f(\overline{u}_i)\,\Delta u_i. \tag{10.14}$$

Summing each area represents the total approximated area $\overline{A}$ under $f(u)$:

$$\overline{A} = f(\overline{u}_1)\,\Delta u_1 + f(\overline{u}_2)\,\Delta u_2 + f(\overline{u}_3)\,\Delta u_3 + f(\overline{u}_4)\,\Delta u_4, \tag{10.15}$$

or

$$\overline{A} = \sum_{u=0}^{4} A_i = \sum_{u=0}^{4} \overline{u}_i\,\Delta u_i. \tag{10.16}$$

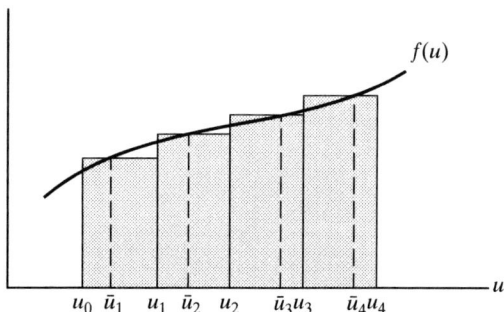

Δu_i	$f(\overline{u}_i)$	$A_i = f(\overline{u}_i)\Delta u_i$
$\Delta u_1 = u_1 - u_0$	$f(\overline{u}_1)$	$f(\overline{u}_1)\Delta u_1$
$\Delta u_2 = u_2 - u_1$	$f(\overline{u}_2)$	$f(\overline{u}_2)\Delta u_2$
$\Delta u_3 = u_3 - u_2$	$f(\overline{u}_3)$	$f(\overline{u}_3)\Delta u_3$
$\Delta u_4 = u_4 - u_3$	$f(\overline{u}_4)$	$f(\overline{u}_4)\Delta u_4$

Figure 10.4   Approximated Area

You can improve your results by shrinking each Δu_i to an "itty-bitty," or *infinitesimal*, size. The limit as $\Delta u_i \to 0$ yields

$$A = \lim_{\Delta u_i \to 0} \sum_{i=1}^{n} f(\bar{u}_i)\,\Delta u_i, \qquad (10.17)$$

where A provides the exact, total area. Equation (10.17) defines an **integral**, which is the limit of a sum of infinitesimal "slices". In general, an integral measures area below a curve to the axis of the independent variable.

10.6.2 Notation

Assume that you have defined a function $y = f(x)$. Let x and y represent the independent and dependent variables, respectively. Borrowing from the notation used for differentiation, you may denote infinitesimal changes in a variable x as dx. Now, consider the following implications:

- The width of each infinitely slim slice is the term dx.
- The height of each infinitely slim slice is the function value $f(x)$.
- The area of each infinitely slim slice is the product $f(x)dx$.

Between end points, there exist an infinite number of slim slices of width dx and height $f(x)$. The total area between end points $x = a$ and $x = b$ must be the sum of all segments of area $f(x)dx$. Calculus denotes this sum with an integral in the form

$$\textbf{\textit{Integral}}: \int_a^b f(x)\,dx, \qquad (10.18)$$

where $\int$ is the integral sign. The integral sign resembles a funny-looking "S," which stands for *sum*.

10.6.3 Physical Interpretation

Integrals help measure cumulative effects of changes, or how much a model varies. Consider the definition of *work = force × displacement*. In integral form, this definition is $W = \int p\,du$. Using the example of a spring, the load p pulls the spring a displacement u. Assuming no energy due to friction, loss—for example—the spring stores all of the work done by the load p as potential energy:

$$E_s = \frac{1}{2}ku^2. \qquad (10.19)$$

Releasing the spring would cause the spring to release stored energy by snapping back.

By slowly pulling the spring, one would cause the load and displacement to gradually increase, according to Hooke's law, $p = ku$. Each increment of load and displacement stores more energy, which you may represent as

$$\Delta E_s = \Delta p \Delta u. \qquad (10.20)$$

These energy increments contribute to the total energy stored inside the spring. Moreover, each $\Delta p \Delta u$ slices the area under $p = ku$ into boxes. An exact sum of the boxes is the total area, or the following integral for E_s:

$$E_s = \int p\,du = \frac{1}{2}\,ku^2. \tag{10.21}$$

10.6.4 Definite Integration

A **definite integral** sums the area below a curve between two points along the independent variable **var**. To find the integral of a function **f** between two points **a** and **b**, enter **int(f, var=a..b)**:

Step 268: Definite Integration

> **int(2*u, u=0..4);**   *Evaluate $\int_0^4 2u\,du$.*

                 16   *Maple finds the area below u^2 between 0 and 4.*

Consult **?int** for more information.

10.6.5 Indefinite Integration

An **indefinite integral** has no specified bounds. This integral produces functions that compute the area below a curve for any bounds. Enter **int(expr,var)** to generate an indefinite integral:

Step 269: Indefinite Integration

> **int(k*u,u);**   *Evaluate $\int ku\,du$.*

           $\frac{1}{2}ku^2$ *This integral produces the area of the triangle below ku.*

Maple does not add a generic constant to results, although the addition of this constant is common practice in hand calculations.

10.6.6 Multiple Integrals

When integrating multivariable functions like $f(x, y)$, use multiple integrals. For instance, enter **Doubleint(f(x,y),x,y)** to evaluate $\int\int f(x, y)\,dx\,dy$. See **?VectorCalculus** for an extensive variety of related functions.

10.6.7 Miscellaneous

For more information about related functions, consult Mathematics…Calculus…Integration… inside the Help Browser. See also the calculus examples in **?examples[index]**.

PRACTICE!

14. Integrate $\sin(x)$ along $0 \leq x \leq \frac{\pi}{2}$.
15. Determine the indefinite integral of $\sin(x)$.
16. Compute the total area *above* and *below* the x-axis for $\sin(x)$ along $0 \leq x \leq 2\pi$.

10.7 FUNDAMENTAL THEOREM OF CALCULUS

This section shows how derivatives and integrals represent inverse operations.

10.7.1 Force and Work

Consider the equations for the spring example. Force is described as $p = ku$. Spring energy is $E_s = \frac{1}{2}ku^2$, and in previous steps, you found that

- The derivative of E_s is $\frac{dE_s}{du} = ku$. Thus, differentiating energy (and work) yields force.
- The integral of p is $\int ku\,du = \frac{1}{2}ku^2$. Thus, integrating force yields work.

The general formulas $W = \int p\,du$ and $p = dW/du$ relate work and force. Integration and differentiation are indeed inverse operations for this example!

10.7.2 Antiderivatives

In general, differentiating an indefinite integral of a function $f(x)$ produces the original function $f(x)$. Thus, indefinite integrals are sometimes called antiderivatives. Denote the antiderivative of $f(x)$ as $F(x)$. Given a continuous function $f(x)$ along $a \le x \le b$, the **Fundamental Theorem of Calculus** (FTC) states:

$$\int_a^b f(x)\,dx = F(b) - F(a). \tag{10.22}$$

$F(x)$ represents the function determined by indefinite integral $\int f(x)\,dx$.

PRACTICE!

17. Demonstrate the FTC with k and $p = ku$ for the spring example that this chapter employs.
18. Demonstrate the FTC with $f(x) = \sin(x)$ for $0 \le x \le 2\pi$.
19. Use the **AntiderivativePlot** function to demonstrate the FTC for $\sin(x)$.

10.8 APPLICATION: DIFFERENTIAL EQUATIONS

Maple provides an excellent tool for solving calculus-based models. This section introduces how calculus provides a tool for modeling dynamic systems with differential equations.

10.8.1 Background

Many branches of engineering, science, and other fields model changing systems with *differential equations*, which are equations that include differentials as parameters. One common differential equation models systems that grow or decay in proportion to their present amount:

$$\text{rate of } Q(t) \propto Q(t). \tag{10.23}$$

Given a quantity $Q(t)$ at time t, the derivative $Q(t)$ with respect to t models the rate of change of $Q(t)$. Thus, you may represent a changing quantity as $dQ(t)/dt$. Because $Q(t)$

changes in proportion to the current amount, you can state the system's governing model as

$$\frac{dQ(t)}{dt} = kQ(t). \tag{10.24}$$

Equation (10.24) assumes a linear relationship represented by the growth rate k. Note that Eq. (10.24) is an *ordinary differential equation* (ODE), which is an equation expressed in terms of one independent variable t.

10.8.2 Problem

Let $Q(0) = Q_0$. Solve (Eq. 10.24).

10.8.3 Methodology

First, state and assign pertinent variables:

Step 270: Initialize Maple

```
> restart:
```
Restart your Maple session.

Step 271: Restate

$Q(t)$ = quantity that changes as a function of time t *Enter text.*

k = parameter (check how it affects the model) *Enter text.*

You can solve "separable" ODEs by rearranging the dependent and independent variables:

- First, divide the entire equation by $Q(t)$.
- Next, multiply both by the differential dt.

You will obtain the following differential equation:

$$\frac{dQ(t)}{Q(t)} = kdt. \tag{10.25}$$

You can express this equivalent ODE in Maple. Start by assigning the left hand side of Eq. 10.25:

Step 272: Separate the LHS

```
> LHS := diff(Q(t),t)/Q(t);
```
Assign $\frac{dQ(t)}{Q(t)}$ to LHS.

$$LHS := \frac{\frac{d}{dt}Q(t)}{Q(t)}$$

Note that older versions of Maple used the partial-differential symbol ∂ instead of d. You still need to assign the right-hand side of Eq. (10.25). You should not include the differential dt! The following step will demonstrate why:

Step 273: Separate the RHS

```
> RHS := k;
```
Assign k to RHS.
$$k$$
Do not include dt.

Recall that integrals reverse differentiation. Therefore, to solve the equation, simultaneously integrate both sides of Eq. (10.25):

$$\int_0^T \frac{dQ(t)}{Q(t)} = \int_0^T k \, dt. \tag{10.26}$$

Do you see how the RHS includes dt as part of the integral? Both integrals use limits of integration $t_{initial} = 0$ and $t_{final} = T$, as indicated by the problem statement. Now, integrate with Maple's **int(expr, var=range)** function:

Step 274: Model

```
> Temp := int(LHS,t=0..T)=int(RHS,t=0..T);
```
Integrate both sides.
$$Temp := \ln(Q(T)) - \ln Q(0) = kt$$
You will need to simplify this result.

Why does Maple report a natural logarithm? Maple knows the integral $\int (1/x)dx = \ln x$. Now, use **solve** to derive an expression for $Q(T)$:

Step 275: Solve and Report

```
> Temp2 := solve(Temp,Q(T));
```
*Try **solve** to rearrange your result.*
$$Temp\,2 := \frac{Q(0)}{e^{-kt}}$$
Well, you still have a little more work to do.

Note that $e^{kt} = \exp(kt)$.

10.8.4 Solution

As the last step, substitute Q_o for $Q(0)$, which is the model's initial condition. The solution to Eq. (10.24) results in an exponential function:

Step 276: Done!

```
> Model := simplify
(subs(Q(0)=Qo,Temp2));
```
Substitute the initial condition Q_o, and simplify the results.
$$Model := Qoe^{kt}$$
Congratulations! You solved a differential equation.

KEY TERMS

area	calculus	continuous functions
definite integral	derivative	first derivative
Fundamental Theorem of Calculus	high-order derivatives	indefinite integrals
integral	limits	partial derivatives
slope	smooth functions	tangent

SUMMARY

- Calculus is the mathematics of change.
- Use the **Student[Calculus1]** library package to learn about calculus in a step-by-step fashion.
- Use **limit** to find the value of a function as it approaches a point.
- For a 2-D function, slope provides a measure of how the function changes at a point, assuming continuity and smoothness there.
- You may use **Tangent** to find the slope of a function.
- A derivative measures the degree of change in a function as its independent variable or variables change, which generalizes the concept of slope.
- Use **diff** to find a numerical and symbolic derivative.
- You can approximate the area below a curve by slicing that area into rectangles and other geometric objects with areas that are easy to determine.
- An integral measures the accumulation of a function as its independent variable(s) change, which generalizes the concept of area.
- Use **int** to find a numerical and symbolic integral.
- Differentiation and integration are inverse operations.

Problems

1. Under what conditions can you take the derivative of a function?
2. Describe how the processes of differentiation and integration relate.
3. For each of the following expressions, use Maple to find the first derivative with respect to the variable x (Hint: Review **?diff**):

    (a) x

    (b) cx

    (c) $cx^2 + x$

    (d) $\sqrt{x}$

    (e) $x + \dfrac{1}{x}$

    (f) xy.

4. Use Maple to evaluate each of the following derivatives. Hint: Review **?diff**:

    (a) Let $y = x^2 + 2x$. Find $\dfrac{dy}{dx}$.

    (b) Let $y = x^2 + 2x$. Find $\dfrac{d^2y}{dx^2}$.

    (c) Let $y = e^x$. Find $\dfrac{d^2y}{dx^2}$. Hint: Recall that e is the exponential function.

    (d) Let $y = \sin(2x)$. Find $\dfrac{d^3y}{dx^3}$.

    (e) Let $z = xy$. Find $\dfrac{\partial z}{\partial x}$.

    (f) Let $z = xy$. Find $\dfrac{\partial z^2}{\partial x \partial y}$.

5.   Use Maple to evaluate the following indefinite integrals:

(a) $\int x\,dx$

(b) $\int x^2\,dx$

(c) $\int \dfrac{1}{x}\,dx$

(d) $\int x \sin x\,dx$

6.   Use Maple to evaluate the following definite integrals:

(a) $\int_0^2 x\,dx$

(b) $\int_0^\pi \cos x\,dx$

(c) $\int_1^2 \int_0^2 x\,y\,dx\,dy$ (Hint: This is called a *double-integral.*)

(d) $\int_{-\infty}^\infty e^{(-x^2/2)}\,dx$ (Hint: Use the exponential function.)

7.   Given a displacement $u(t)$ and time t, the derivative $\dot{u}(t)$ computes velocity, and the derivative $\ddot{u}(t)$ computes acceleration. Let $u(t) = t^2 + 4t - 6$ m. Perform the following steps using Maple:

(a) Assign the variable u to the function $t^2 + 4t - 6$.

(b) What kind of units would you normally expect the displacement u to have?

(c) Find the velocity $\dot{u}$ by evaluating the derivative of u with respect to time t. Assign the name v to your results. Hint: Use **diff(something,t)**.

(d) What kind of units would you normally expect the velocity v to have?

(e) Find the acceleration $\ddot{u}$ by evaluating the derivative of $\dot{u}$ with respect to time t. Assign the name a to your results. Hints: Use **diff(something,t)** or **diff(u,something)**.

(f) What kind of units would you normally expect the acceleration a to have?

8.   Confirm your results in Problem 7. Note that you must use the *initial conditions* $u(0) = -6$ and $\dot{u}(0) = 4$ to properly check your expressions:

(a) Starting with acceleration, integrate $\ddot{u}$ with respect to time t. Hint: Try **int(a,t)**. You will find an expression for v. Now, modify v with the proper initial condition.

(b) Next, integrate $\dot{u}$ with respect to time t. Hint: You will find an expression for u. Now, modify u with the proper initial condition.

(c) Do your results match the equations in Problem 7?

9.   Consider the *electric field E*, the force per unit charge, that the charged rod in Figure P10.5 exerts on any point P along the rod's axis. From physics, assuming a charge per unit length of λ,

$$\Delta E = \frac{k\Delta q}{x^2} = \frac{k\lambda\Delta x}{x^2} \qquad (10.27)$$

given unit charge Δq.

To derive an equation for E, let ΔE and Δx shrink to infinitesimal sizes, dE and dx, respectively. Thus,

$$dE = \frac{k\lambda\,dx}{x^2} \qquad (10.28)$$

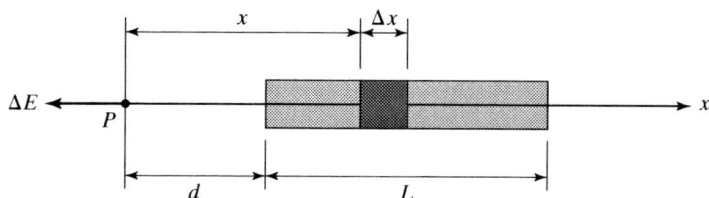

Figure P10.5 Electric Field of Charged Rod

Using Maple, find an equation for E using the geometry indicated in Figure P10.5. You should obtain the result that

$$E = \frac{k\lambda L}{(d + L)d}.$$

Hints: Enter **int(1,E)** to prove that integrating dE yields just E. You now need to solve a definite integral for $k\lambda\, dx/x^2$ along the range $d \leq x \leq d + L$.

10. The Clapeyron equation relates a changing pressure P and temperature T of a substance. An ordinary differential equation expresses a simplified form of the relationship

$$\frac{dP}{P} = \frac{h_g - h_0}{R}\frac{dT}{T^2} \tag{10.29}$$

for the constants h_g, h_0, and R. Given initial values (P_1 and T_1) and final values (P_2 and T_2), solve this equation. Hints: Follow the steps in Section 10.8. You might also wish to try **dsolve**.

11. As shown in Figure P10.6, the *bending moment* M_0, with units of *force · length*, twists the cantilever beam in the xy-plane. Bending moments are defined as forces that act upon bodies and cause twisting about some axis. The differential equation

$$\frac{d^2y}{dx} = \frac{M}{EI} \tag{10.30}$$

relates the beam's deflection y with respect to the distance x, the bending moment M, and the flexural stiffness EI. Assuming constant EI, the solution to the differential equation yields

$$y = \frac{M_0 x^2}{2EI}, \tag{10.31}$$

where both M and y are functions of x.

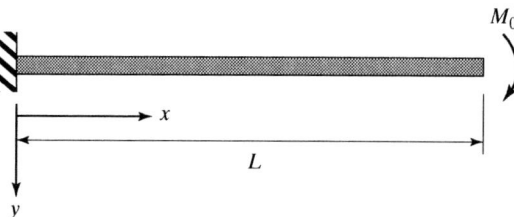

Figure P10.6 Cantilever Beam

Answer the following questions using Maple:

(a) Solve for the bending moment as a function of x. Hint: Use the equation for y and the differential equation. You will need a second derivative.

(b) Plot both the moment M and the displacement y as a function of x on the same graph.

(c) Does the moment change as a function of position x?

(d) What is the shape of the deflection curve? Hint: Consider the nature of the independent variable x.

(e) Does the top of the beam experience compression or tension? Why or why not?

12. Repeat the problem in Section 10.8 for these values: $k = 2$ and $k = -2$. Explain the difference between choosing positive and negative values of k. You might also wish to supply plots that illustrate your answer.

13. Demonstrate how spring force $p = ku$ and spring energy $E_s = \frac{1}{2}ku^2$ are related. Hint: Use the FTC.

14. Consider the motion of a mass connected to a spring, as shown in Figure 10.7. See also Figure P10.7. The mass will bounce up and down after the hand releases the mass.

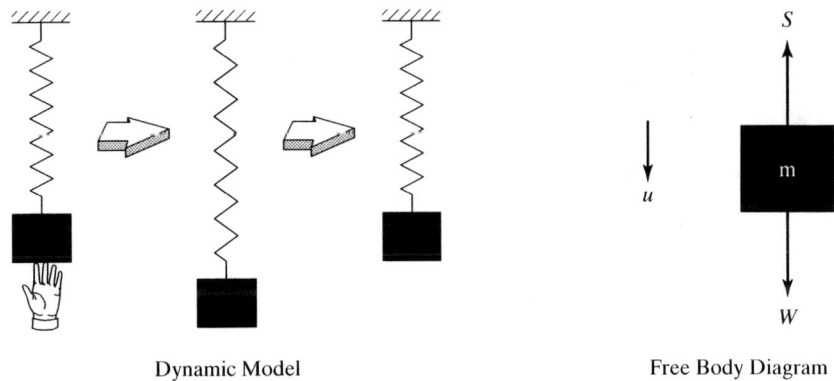

Dynamic Model   Free Body Diagram

Figure P10.7   Mass-Spring Model

Whereas gravity initially pulls the mass downward, the spring reacts by pulling upward. Newton's second law states that the *rate of change* of momentum $m\,du/dt$ experienced by a mass equals the force F_u acting on the mass m. Thus,

$$\sum F_u = ma_u, \tag{10.32}$$

where acceleration

$$a_u = \frac{d^2u}{dt^2} = \ddot{u}, \tag{10.33}$$

given displacement u and constant mass. Solve the following problems to describe the motion of the mass:

(a) Express a_u in Maple as a function of displacement $u(t)$. Big hint: Restart Maple, and then, enter **a:=diff(u,t,t)**.

(b) From Newton's Second Law and the free-body diagram as shown in Figure 10.7, $W - S = m\ddot{u}$. Using your expression for $\ddot{u}$, assign this equation to the name NSL. Why should you use the name $Weight$ instead of W? Maple should output the following:

$$NSL: = Weight - S = m\left(\frac{d^2}{dt}u(t)\right). \tag{10.34}$$

Using Hooke's law and spring stiffness k, the weight provides an initial static displacement u_0. Thus, you can relate the weight and static displacement:

$$W = ku_0. \tag{10.35}$$

Substitute this equation for weight into your expression for NSL. Assign the result to $NSL2$. Maple should output the following:

$$NSL2: = ku0 - S = m\left(\frac{d^2}{dt}u(t)\right). \tag{10.36}$$

The total displacement that the mass experiences at any time t is the sum $u(t) + u_0$. You may now express the spring force S in terms of k and displacements $u(t)$ and total displacement of the mass as

$$S = k(u(t) + u_0). \tag{10.37}$$

Substitute this equation for weight into your expression for $NSL2$. Assign the results to $NSL3$. Maple should output the following:

$$NSL3: = ku0 - k(u(t) + u0)S = m\left(\frac{d^2}{dt^2}u(t)\right). \tag{10.38}$$

(c) On the basis of expression $NSL3$, show that $m\ddot{u} + ku = 0$ using Maple. Maple should output the following:

$$ODE: = -ku(t) = m\left(\frac{d^2}{dt^2}u(t)\right).$$

Based on these results, discuss whether initial static displacement affects the system's model.

(d) Express ODE as $ODE2: = -u(t)\omega^2 = \frac{d^2}{dt^2}u(t)$. Hints: Substitute $\omega^2 = \frac{k}{m}$ with an algebraic substitution into $(ODE)/m$. Also, that's an omega, not a "w"!

(e) Solve your differential equation using **dsolve**. Assign the results to U. You should obtain this result:

$$U: = u(t) = \_C1\cos(\omega t) + \_C2\sin(\omega t). \tag{10.39}$$

(Maple might swap the coefficients $\_C1$ and $\_C2$.)

(f)   Assign the velocity $v(t)$ to V, which is defined as $\dot{u}(t)$. Hint: Use **diff** on the expression U. You should obtain the following:

$$V: = \frac{d}{dt} u(t) = \_C1 \cos(\omega t) + \_C2 \sin(\omega t). \qquad (10.40)$$

(g)   Let displacement $u(0) = U_0$ and velocity $\dot{u}(0) = v(0) = V_0$. Express U and V in terms of U_0 (**Uo**) and V_0 (**Vo**). Hints: You need to solve for the coefficients $\_C1$ and $\_C2$. Separate out the RHS of U and V. Evaluate the expressions for $t = 0$. You might wish to try **eval** and **solve**. You should obtain these results:

$$u(t) = Uo \cos(\omega t) + \frac{Vo \sin(\omega t)}{\omega} \qquad (10.41)$$

and

$$\frac{d}{dt} u(t) = -Uo \sin(\omega t)\omega + Vo \cos(\omega t). \qquad (10.42)$$

(h)   Let $U_0 = 1$, $V_0 = 0$, and $m = k = 1$. Plot both $u(t)$ and $v(t)$ on the same graph for $0 \le t \le 2\pi$. Be sure to label your plots.

(i)   What is the maximum displacement of the mass?

(j)   What is the velocity whenever the mass reaches its maximum displacement?

Appendix A

Symbols

A.1 KEYBOARD CHARACTERS

!	Exclamation Point, *Bang*	(	Left Parenthesis, *Open Paren*
@	At Sign	)	Right Parenthesis, *Closed Paren*
#	Pound Sign, *Sharp, Hash*	[	Left Square Bracket
$	Dollar Sign	]	Right Square Bracket
%	Percent Sign	{	Left Brace Bracket, *Left Curly Brace*
∧	Circumflex, Caret	}	Right Brace Bracket, *Right Curly Brace*
&	Ampersand, *And*	<	Left Angle Bracket, *Less Than*
*	Asterisk, *Star*	>	Right Angle Bracket, *Greater Than*
+	Plus Sign, *Add*	/	Forward Slash, Virgule, *Slash*
=	Equal Sign	\	Backslash, *Switch*
~	Tilde	:	Colon
?	Question Mark	;	Semicolon
|	Vertical Line, *Bar, Pipe*	`	Back Quotation Mark, Grave Accent
_	Underscore, Underline	'	Single Quotation Mark, Apostrophe
-	Hyphen, Minus Sign, *Dash*	"	Double Quotation Mark
.	Period, *Dot*	,	Comma

A.2 GREEK CHARACTERS[1]

A	Alpha	α	alpha
B	Beta	β	beta
Γ	Gamma, GAMMA	γ	gamma
Δ	Delta	δ	delta
E	Epsilon	ε	epsilon
Z	ZETA	ζ	Zeta, zeta
H	Eta	η	eta
Θ	Theta	θ	theta
I	Iota	ι	iota
K	Kappa	κ	kappa
Λ	Lambda	λ	lambda
M	Mu	μ	mu
N	Nu	ν	nu
Ξ	Xi	ξ	xi
O	Omicron	o	omicron
Π	PI	π	Pi, pi
P	Rho	ρ	rho
Σ	Sigma	σ	sigma
T	Tau	τ	tau
Y	Upsilon	υ	upsilon
Φ	Phi	ϕ	phi
X	CHI	χ	chi
Ψ	Psi	ψ	psi
Ω	Omega	ω	omega

---

[1]Beware of names such as `Pi` and `gamma` that have predefined values. Enter `type(name, protected)` to check. Also, see `?greek` and `?symbolfont`.

Appendix B

Maple Functions

Maple stores a function essentially in one of two locations: the **kernel** or the **library**. The library has two categories: the **main library** and the **library packages**. This appendix briefly discusses these categories, and you can access functions inside each.

B.1 KERNEL

Maple's kernel is a core collection of functions that are built-in, which means you cannot see their code. If you are curious which functions belong to the kernel, try the following command:

```
> select(type,{unames(),anames(anything)},builtin);
```
A rather lengthy set of built-in functions will appear.

B.2 MAIN LIBRARY

Other than kernel commands, Maple's **main library** contains the bulk of commonly used functions. Calling one of these functions in an input statement automatically loads the function into memory. If the function's Help file does not indicate any special loading procedure, the function belongs to the main library.

B.3 LIBRARY PACKAGES

Library packages contain functions for specialized forms of analysis. Examples of library packages used in this text include **plots** and **LinearAlgebra**. Library packages contain functions that Maple does not automatically recognize as belonging to the main library. Thus, you provide different commands to access the functions contained inside those packages. Why doesn't Maple just automatically load the library packages? Many functions that are stored within the packages have names that clash with functions that are stored within other packages. Also, users might prefer to use names that functions within library packages already use. Consult **?index[package]** for more information on the available library packages. You should also investigate the Maple Application Center (see **?webresources**) for other applications.

B.4 ACCESSING FUNCTIONS FROM LIBRARY PACKAGES

This section discusses three methods for accessing functions from the library packages. See **?with** for more information.

B.4.1 Method 1

Access the ***long name***, which means that you indicate the package and command names in one expression with the syntax ***package[function](args)***. For instance,

```
> storedPlot := plot(2*x,x):                    Store a plot structure.
> plots[display](storedPlot);                   Display the stored plot.
```

If the package has a subpackage, use ***package[subpackage][function](args)***:

```
> Student[Calculus1][ApproximateInt](2*x,x=0..4);   Find area below 2x.
```

You may discover some oddities in your explorations, as in the **stats** package, which requires the syntax ***package[subpackage, function](args)***:

```
> stats[ fit, leastsquare[[x,y]] ] ([[1,2,3],[2,4,6]]);
```

B.4.2 Method 2

Access the ***short name*** of a single command in a package. A short name is the name of the function without any of the package/subpackage information. First, you must enter **with(*package, function*)** or **with(*package[subpackage], function*)**, depending on function's location. Then, you may use the short name function by itself:

```
> with(plots,display);                          Display the stored plot.
> display(plot(2*x,x));                          Display the stored plot.
> with(Student[Calculus1],ApproximateInt);      Find area below 2x.
> ApproximateInt(2*x,x=0..4);                    Find area below 2x.
```

B.4.3 Method 3

Access all short names of all functions in a given package. Actually, I tend to use this technique most often because it's the easiest to type, and I get lazy. Use the syntax **with(*package*)** or **with(*package[subpackage]*)**. Afterwards, you may use any commands in the package/subpackage until you restart:

```
> with(plots);                                  Display the stored plot.
Warning, the name changecoords                  This is NOT an error message!
has been redefined
```
Maple prints a list of functions that you may now access, which I am not showing.

```
> display(plot(2*x,x));                          Display the stored plot.
```

```
[> with(Student[Calculus1]);                    Find area below 2x.

[> ApproximateInt(2*x,x=0..4);                   Find area below 2x.
```

What is the warning that entering **with(plots)** causes? Some function names "collide" with standard Maple Function names, which sometimes necessitates using Method 1 when you do not want this conflict. You may also wish to use the colon (**:**) to suppress the printing of all functions to which you obtain access. If you lose track of all the packages you have made accessible, enter **packages()** to see the full list. Consult **?with** for more information.

B.5   FINDING FUNCTIONS

Investigate the appropriate Help pages for information on the following functions:

- Main library: **?inifcn** and **?index[function]**.
- Library packages: **?index[package]**.
- Applications: **?webresources**.

B.6   VIEWING FUNCTIONS

Except for built-in commands, you can view all of the actual code that Maple uses by setting the interface variable called **verboseproc**:

```
[> interface(verboseproc=2);          Tell Maple to report as much as possible.
```

To see the code that Maple actually uses, enter **print(name)**:

```
[> print(sin);                        Display the code for the sin function.
[ Very long output...
```

Set **verboseproc** to **0** to return to Maple's normal mode of operation, which abbreviates the display for functions:

```
[> interface(verboseproc=0);          Make Maple's reporting mode sparse again.

[> print(sin);                        Try to display the code for sin.
[        proc (x::algebraic) ... end   Maple reports an abbreviated message.
```

Consult **?interface** and **?print** for more information.

Appendix C

Scientific Constants and Units

Maple offers a tremendous resource if you need to solve science and engineering problems, as I hope that this text has demonstrated. There are two relatively new library packages that assist with such applications: **Units** and **ScientificConstants**. In fact, Maple now organizes these packages in the Help Browser under Science and Engineering..., likely in anticipation of future additions. Consult Appendix B for information on how to access the functions in these packages.

C.1   UNITS

This section gives an overview of the Units package, which you will find described in **?Units**.

C.1.1   Understanding Units

Science and engineering tend to focus on understanding the world around us, which involves measuring and predicting different states and their changing values. We measure physical quantities in units. For instance, I sometimes like to ask students why 2 + 2 might equal 1 instead of 4. The real answer depends on whether or not I am talking about 2 things plus another 2 things, all of which are the same things–who knows? Supplying units along with the numbers would greatly clarify the problem!

Maple uses some terms with which you might not be familiar. See the following definitions, which are adapted from **?Units[Details]**:

- **Base dimension**: measurable quantity that units are composed of, such as length, time, and electric current. For a full list, enter **GetDimensions (base=true)**.
- **Base unit**: a unit that corresponds to a base dimension, like meter, second, and ampere.
- **Derived unit**: a named unit that is composed of products of powers of other units, as in volume or force.

For a full list of units that Maple understands, see **?Units[Overview]** and Science and Engineering...Units...Known Units....

Maple also uses ***systems of units***, which are collections of base and derived units. You probably have heard about *SI*, or The International System of Units (see **?Units[SI]**). You can enter **Units[GetSystems]()** to see all of the systems that Maple currently supports. See **?Units[GetSystem]**, **?Units[Systems]**, and **?Units[Details]** for more details.

C.1.2 Conversions

If you are struggling with converting units to other units, use Maple! It knows about almost all unit conversions out there. You can use Maple for basic conversions with **convert**, which has one form of syntax as **convert(*expr*, *units*, *UnitsFrom*, *UnitsTo*)**. For instance, you can perform operations such as finding out how many centimeters there are in an inch:

> **convert(1,units,inch,cm);**   *Convert in to cm.*
$$2.540000000$$ *Actually, there are exactly* 2.54 *cm/in.*

> **convert(1,units,furlong/fortnight,m/s);**   *Paul Revere for a modern age!*
$$\frac{1397}{8400000}$$ *Use* **1.0** *or* **evalf** *for a numerical answer.*

Note that you can use any ***expr*** instead of just **1** in the preceding examples. You can also have Maple explain the base and complex dimensions of a unit. The following unit name always amused me back when I took introductory physics:

> **convert(m/s^3,dimensions);**   *You can make Maple "insult you."*
$$jerk$$ *Actually, jerk is change of acceleration per unit time.*

For more details and options, see **?convert[units]** and Science and Engineering... Units...Conversions.... for other options, like **?convert[dimensions]**. For information on a conversion subpackage of **Units**, see **?Units[Converter]** and Edit→Unit Converter....

C.1.3 Environments

Maple has three environments in which it handles units, as discussed in Science and Engineering...Units...Environments...:

- ***Default environment* (?Units[Default])**: You cannot directly use units, except in some conversion routines.
- ***Standard environment* (?Units[Standard])**: You can use units in some expressions with a special function that is called **Unit**.
- ***Natural environment* (?Units[Natural])**: You can use units in some expressions directly with their names or the **Unit** function.

For standard and natural environments, the Help windows give lists of expressions and functions that accept units. I recommend the natural environment, unless you tend to confuse unit names with variables. When using units to solve a problem, I also recommend that you set a system of units, like the "English System" (FPS: foot-pound-second) or the

"Metric System" (SI: International System). In the following examples, I demonstrate how to use a system of units in the the natural environment, and then I convert the answer to a different set of units:

> `Units[UseSystem](FPS):`   *Instruct Maple to use English units.*

> `with(Units[Natural]):`   *Maple will allow you to use unit-labels in your input.*
Ignore Maple's greatly-detailed warning about the functions it has modified.

> `momentum := 150*lbs * 60*mph;`   *Find the momentum of a person in your car on the highway.*

$$momentum := 880 \left[\frac{lb\,ft}{s} \right]$$ *This number explains the need for traffic safety!*

> `evalf(convert(momentum,system,SI),4);`   *Convert to SI units.*

$$121.7 \left[\frac{kg\,m}{s} \right]$$ *Eventually, the USA will switch to SI.*

I recommend that you investigate the many examples that Maple has provided in Science and Engineering...Units...Worksheets....

C.2 SCIENTIFIC CONSTANTS

Maple provides a multitude of scientific constants, even including the entire Periodic Table! See the following sources of help for an overview of the features:

- Overview of package: Science and Engineering...Scientific Constants... Introduction or **?ScientificConstants**.
- Classification of functions: Science and Engineering...Scientific Constants... Details.
- Example worksheets: **?SCApps**.

This section gives a brief overview of the main features to get you started.

C.2.1 Constants

Throughout your studies you will encounter many constants, such as the standard acceleration of gravity (g) and speed of light in a vacuum (c). You can find Maple's built-in collection at **?ScientificConstants[constants]**. For a complete list of functions that you can use, see **?ScientificConstants[Physical]**. To supply a constant in an expression, you need to ensure that the **ScientificConstants** package is accessed.

In the following example, I access the gravitational acceleration constant $g = 9.81$ m/s^2 with the function **Constant**. By specifying **Constant(g,units)**, Maple will retrieve the built-in value of g with its units in the current unit system, which defaults to SI. I also force Maple to produce a numerical answer with **evalf**. In the following example, I determine how much mass a 200-pound person has in terms of kilograms:

```
> with(Units[Natural]):
```
Maple will allow you to use unit-labels in your input.
Ignore Maple's greatly-detailed warning about the functions it has modified.

```
> with(ScientificConstants):
```
Access Maple's built-in constants.

```
> g_SI := evalf(Constant(g,units));
```
Access and evaluate g.

$$g\_SI := 9.80665 \left[\frac{m}{s^2}\right]$$ *Maple gives units because of the* **units** *option.*

```
> W := 200*lbf;
```
lbf means pound-force, as opposed to mass.

$$W := 200[lbf]$$ *Maple does not use SI because you used FPS here.*

```
> Mass := evalf(W/g_SI,2);
```
Find the mass in terms of default units.

$$Mass := 89.[kg]$$ *m=W/g. Maple uses SI, which is the default system.*

For more information, see **?Constant**.

C.2.2 Elements

The **ScientificConstants** package has the Periodic Table and properties of all the elements. See **?ScienfificConstants[Periodic]** for an overview of the entire functionality. To access an element and its information, use **GetElement**, as in **GetElement(gold)**, for instance. Out of curiosity, you can see how much a room $(10 \times 10 \times 2\ \mathrm{m}^3)$ *completely* full of pure gold actually weighs:

```
> restart:
```
Clear all packages and assignments.

```
> with(Units[Natural]):
```
Maple will allow you to use unit-labels in your input.
Ignore Maple's greatly-detailed warning about the functions it has modified.

```
> with(ScientificConstants):
```
Access Maple's built-in constants.

```
> rho[Au] := evalf(Element(gold,density,units));
```
Density of Gold (Au).

$$\rho_{Au} := 19300.0 \left[\frac{kg}{m^3}\right]$$ *Maple reports SI units, which are default.*

```
> Volume := (10*m)*(10*m)*(2*m);
```
Volume of a $10 \times 10 \times 2\ \mathrm{m}^3$ room.

$$Volume := 200[m^3]$$ *Maple combines the length units.*

```
> g_SI := evalf(Constant(g,units));
```
Access and evaluate g.

$$g\_SI := 9.80665 \left[\frac{m}{s^2}\right]$$ *See* **?Constant***.*

```
> Weight := evalf(rho[Au]*Volume*g_SI,3);
```
W=mg=ρVg.

$$1.64\ 10^8 [N]$$ *That's a lot of gold!*

If you prefer pounds, enter **convert(Weight,units,N,lbf)/1e6**, and you will see that a "room full of gold" weighs over 8 million pounds!

Appendix D

Introduction to Programming

This appendix briefly introduces aspects of programming in Maple. Consult **Programming...** and more advanced Maple books for intensive discussion of Maple's programming features and techniques.

D.1 VARIABLES

Recall that assignable names are *variables*, terms that can store different expressions. Maple employs three kinds of variables:

- **Global Variables**: assignable names known to all Maple functions and procedures. Virtually every name you assign at the prompt is global. Consult **?global** or **?procedure**.
- **Local Variables**: names that exist only within library functions and customized procedures. Local variable values do not change global names. Consult **?local** or **?procedure**.
- **Environment Variables**: predefined constants that affect how certain Maple functions perform. These variables can affect both local and global behavior. Consult **?envvar**.

See **?names** and Programming...Names and Strings... for more information.

D.2 ADVANCED DATA TYPES

Programming in Maple often requires expression types like *boolean*, *relation*, *array*, and *table*. Consult Programming...Data Types... for an extensive list of different types. This section covers **array** and **table**, which were not covered in depth in other sections.

D.2.1 Arrays

An array organizes information into rows and columns much as a spreadsheet does, as shown in Figure D.1. In fact, matrices and vectors are special cases of arrays. Enter arrays with a list of lists, **array([[row1],[row2],...])**, where each row forms a list, **[col1, col2,...]**:

```
> A := array([[a,1,2,b],[a,3,d,a]]);
```
Assign an array of values to A.

$$A := \begin{bmatrix} a & 1 & 2 & b \\ a & 3 & d & a \end{bmatrix}$$
*Consult **?array** for more information.*

```
> A[1,3];
```
Access A_{13}, the element corresponding to row 1 and column 3 inside A.

$$d$$
*Consult **?indices** for more information.*

If you find that your array does not print, you might be encountering **last name evaluation**, which prevents extremely large expressions from printing. If so, use **eval(*array*)** or **print(*array*)**. For more information, consult **?array**, **?Array**, and **?last_name_eval**.

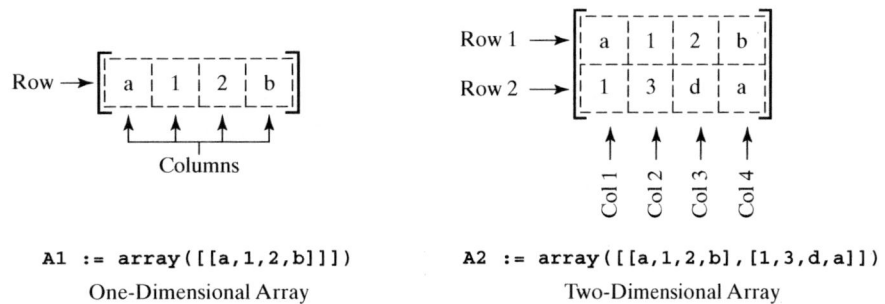

```
A1 := array([[a,1,2,b]])
```
One-Dimensional Array

```
A2 := array([[a,1,2,b],[1,3,d,a]])
```
Two-Dimensional Array

Figure D.1 Array

D.2.2 Tables

Tables resemble arrays, with the exception that table indices can use labels other than integers. In general, enter tables as **table([*index1=value*, *index2=value*,...])**:

```
> person := table([name=Ira,ID=12345]);
```
$person := table([name = Ira, ID = 12345])$
Create a table with two attributes.

To access an entry, use the index label:

```
> person[name];
```
Ira
Access the value corresponding $person_{name}$.
*Consult **?indices** for more information.*

For more information, consult **?table**.

D.3 PROGRAMMING STATEMENTS

Programming involves Maple statements such as **repetition** and **selection**. Investigate **?keyword** for a complete list of Maple's reserved words, most of which Maple uses for programming. You can also find a collection of statement types according to their keywords/commands with **?index[statement]**.

D.3.1 Selection Statements

Selection statements test conditions in order to perform certain tasks. Different cases can then activate other statements:

General Form of Selection Statement:
```
> if expr then statements                    If expr is true, then perform statements.
>      elif expr then statements             Otherwise, if this expr is true, perform these
                                                                              statements.
>      elif expr then statements             Otherwise, if this expr is true, perform these
                                                                              statements.
>      ...
>      else statements                 If no exprs are true, then perform these statements.
> end if;                                             End the selection statement.
```
Maple requires everything except for the `elif expr then statements` *and* `else statements` *portions.*

Use boolean types for each **expr**. If **expr** evaluates as *true*, Maple performs the indicated statements following the **then** keyword. Otherwise, Maple moves to the next **elif** ("else if") condition and tests **expr**. If every **elif** condition fails, Maple activates the statements following the **else** keyword. If all **expr**s fail, Maple takes no action:

Selection Statement Example:
```
> restart:                                                  Clear everything.
```

```
> val := 1:                                             Give val a value of 1.
```

```
> if type(val,even) then print("Even!");       Check if the value is even.
> elif type(val,odd) then print("Odd!");    Otherwise, check if the value is odd.
> else print("What?");                             Otherwise, report an error.
> end if:                                             End selection statement.
            "Odd!"                        Maple outputs the result after you press Enter.
```

As strange as it might seem, you also can use **if** as an operator! Investigate **?if** for more information.

D.3.2 Repetition Statements

When you need to repeat a sequence of operations, use repetition statements. Maple provides two general forms:

(1) General Form of Repetition Statement:
```
> for name from expr by expr to expr while expr
      do statements end do;
```
Maple requires only the portion **do statements end do**. *All other keywords and expressions are optional.*

(2) General Form of Repetition Statement:
```
> for name in expr while expr
      do statements end do;
```
Maple requires only the portion **do statements end do**.

For example, try creating an array of squared integers between 1 and 100:

Repetition Statement Example:
```
> A:=array(1..10);
```
$$A := \text{array}(1..10,[])$$
Create an empty ten element, one-dimensional array.
Maple reports that your array is currently empty.

```
> i:=1:
```
Initialize a counter.

```
> for i from 1 to 10 do
>     A[i]:=i^2:
> end do:
```
Increment the counter **i** *from* 1 *to* 10.
Store $\mathbf{i}^2$ *in* A_i *.*
Repeat the loop until **i** *has no iterations left.*

```
> print(L);
```
$$[1, 4, 9, 16, 25, 36, 49, 64, 81, 100]$$
Need to print because of last name evaluation.
Maple reports the array's contents.

Investigate **?repetition** or **?do** for more information.

D.3.3   File I/O

File I/O consists of reading input into Maple and printing output to a file. For information on how to input and output numerical data, consult both **?readdata** and **?writedata**. The Help Browser describes further commands inside Programming...Input and Output....

D.4   PROCEDURES

As discussed in **?proc**, you can define your own Maple programs that act like Maple functions by using procedures and Maple's **proc** function:

General Form of Procedures:
```
> proc(argseq)
>         local nameseq
>         global nameseq
>         options nameseq
>         description stringseq
>       statements
> end proc;
```
Procedure name and inputs.
Declare local variables, which only **proc** *uses.*
Declare global variables, which all procedures can use.
Modify the behavior of **proc***.*
Label your procedure with inert commentary.
Provide the main body of the procedure.
End the procedure.

The **local**, **global**, **options**, and **description** statements are optional. In the statements with **nameseq**, you would specify a sequence of names. The **statements** portion provides the main body of a procedure with statements separated by standard statement terminators (**:** and **;**). For instance, try creating a procedure that finds the magnitude of a vector:

Procedure Example:
```
> restart:
```
Clear everything.

```
> mag := proc(v)
>         description "Find vector magnitude":
>         local dim, i, temp:
```
Assign a procedure called **mag***.*
Describe procedure.
Declare local variables.

```
>         temp:=0:                              Initialize your sum to zero.
>         dim := LinearAlgebra[Dimension](v):           Size of vector.
>         for i from 1 to dim do                       Start a for loop.
>             temp := temp + v[i]^2:       Sum the squares of each vector entry.
>         end do:                                      End the for loop.
>         sqrt(temp);
> end proc:                                            End the procedure.
```

$$v_{mag} = \sqrt{v_i^2}$$

Now, call your procedure as you would a standard Maple function:

```
> v := <1 | 2 | 3>:                               Assign a row vector to v.
```

```
> mag(v);                              Call the mag procedure to find the magnitude of v.
```
$$\sqrt{14}$$
$$\sqrt{1^2 + 2^2 + 3^2} = \sqrt{1 + 4 + 9} = \sqrt{14}$$

How would you "normally" find a vector magnitude?

```
> with(LinearAlgebra,VectorNorm):        Access Maple's VectorNorm function.
```

```
> VectorNorm(V,Frobenius);        Compare your results with that of VectorNorm.
```
$$\sqrt{14}$$
They're the same!

You could also use **interface(verboseproc=2)** and **print(VectorNorm)** to see Maple's version, which you will find a bit more complicated. However, you now have enough information to start exploring how to write your own Maple functions!

D.5 MODULES

If you are interested in object oriented programming with Maple, I suggest that you check out **?module** and **?examples[obj]**.

Appendix E

Additional Features

This section introduces special features available to some platforms.

E.1 SPREADSHEETS

Maple includes *spreadsheet* capabilities. Besides providing convenient templates for sorting and organizing data, spreadsheets help you apply the same formulas simultaneously to large amounts of data. Consult **?spreadsheet** for a complete overview. Windows users should also see **?excel** to discover how you call Maple from Excel.

E.2 IMPORTING IMAGES

To import external graphical images, select Insert → Object (not available with Unix).

E.3 MATLAB

MATLAB is a popular program for numerical analysis. You call MATLAB routines from within Maple, assuming you have Matlab installed. Consult **?matlab** for instruction.

E.4 WORLD WIDE WEB

You can export worksheets to ***Hypertext Markup Language*** (HTML) format. Moreover, Maple will automatically convert plots and animations. Consult Connectivity...Web Features... for more information.

E.5 MAPLETS

Maple now incorporates Java, which allows you to build user interfaces that allow users to interact with Maple. These programs are called ***Maplets***, which is based on the concept of Java applets for Maple. See **?Maplets** and **?Java** for more information.

Appendix F

Bibliography

F.1 MAPLE INFORMATION

- **http://www.maplesoft.com**

F.2 MATHEMATICS

- **http://e-math.ams.org**
- **http://euclid.math.fsu.edu/Science/math.html**
- **http://www.ams.org/mathscinet**

F.3 MAPLE REFERENCES

Adams, S. G. *Maple Talk*. New Jersey: Prentice Hall, 1997.

Monagan, M.B., K. O Geddes, K. M. Heal, G. Labahn, S. M. Vorkoetter, J. McCaron, P. DeMarco, *Maple 8 Introductory Programming Guide*. Waterloo, ON: Waterloo Maple, 2002.

Nicolaides, R. and N. Walkington. *Maple: A Comprehensive Introduction*. Melbourne, Australia: Cambridge University Press, 1996.

Waterloo Maple, Inc. *Maple 8 Learning Guide*, Waterloo, ON: Waterloo Maple, 2002.

F.4 ENGINEERING, MATHEMATICS, AND SCIENCE REFERENCES

Throughout you education, your professors will advise you to keep your books. As I discovered, you just never know when they will become handy!

American Institute of Steel Construction (AISC). *Torsional Analysis of Steel Members*. AISC: 1983.

American Society of Civil Engineers (ASCE). *Pipeline Design for Hydrocarbon Gases and Liquids*. ASCE: 1975.

Blank, L. T. and A. J. Tarquin. *Engineering Economy*. McGraw-Hill, 1998.

Boyce, W. E. and R. C. DiPrima. *Elementary Differential Equations and Boundary Value Problems*. New York: John Wiley & Sons, 1986.

Das, B. M. *Principles of Foundation Engineering*. Boston, Mass.: PWS-Kent Publishing Company, 1990.

Hwang, N. H. C. and C. E. Hita. *Fundamentals of Hydraulic Engineering Systems*. New Jersey: Prentice Hall, 1987.

Lindeburg, M. R. *Engineer-in-Training Reference Manual*. Belmont, CA: Professional Publications, Inc., 1998.

Paquette, R. J., N. J. Ashford, and P. H. Wright. *Transportation Engineering: Planning and Design*. John Wiley & Sons, Inc., 1982.

Purcell, E J. and D. Varberg. *Calculus with Analytic Geometry*. New Jersey: Prentice Hall, 1984.

Serway, R. A. *Physics for Scientists and Engineers*. New York: Saunders College Publishing, 1986.

Shames, I. H. *Engineering Mechanics*. New Jersey: Prentice Hall, 1990.

Shames, I. H. *Introduction to Solid Mechanics*. New Jersey: Prentice Hall, 1989.

Soong, T. T. *Probabilistic Modeling and Analysis in Science and Engineering*. New York: John Wiley & Sons, 1981.

Spiegel, M. R. *Mathematical Handbook of Formulas and Table*. New York: McGraw-Hill, 1968.

Tabler, R. D. *Design Guidelines for the Control of Blowing and Drifting Snow*. Washington, DC: Strategic Highway Research Program, 1994.

Van Vlack, L. H. *Elements of Materials Science and Engineering*. Addison Wesley, 1985.

Van Wylen, G. J. and R. E. Sonntag. *Fundamentals of Classical Thermodynamics*. New York: John Wiley & Sons, 1985.

Appendix G

Complete Solutions to Practice! Problems

Many practice problems assume or require you to either

- Clear variable assignments by unassignment (**?unassign**) or
- Refresh your worksheet with the **restart** command (**?restart**).

G.1 CHAPTER 1: INTRODUCTION

1.1.   Congratulations! You discovered this appendix.

1.2.   Check with your system administrator.

1.3.   Hopefully at least version Maple 8. In a worksheet, enter **interace(version)** or select the Help menu to check.

1.4.   Use Maple's text search with Help→Full Text Search.... In Word(s), enter **suck** and press Search. You will discover the Help window for **?plottools [vrml]**. Use Edit→Find... in the Help window to find *suck*. Yes, I do find this amusing.

1.5.   Use File→Preferences.... Activating Balloon Help causes Maple to display pop-up windows for icons and menu options.

1.6.   See Getting Started...Using Help...glossary or Reference...glossary. Entering **?glossary** gets you there a bit quicker.

1.7.   Select File→New.

1.8.   Click inside the worksheet and select File→Save As.... You should use the *Maple Worksheet* file-type, which is usually the default setting. The title bar on the worksheet will display the filename that you chose.

1.9.   Select File→Print.... Select the option for printing to a file and give the filename.

1.10.   Select File→Close.

1.11.   Select File→Open.... Choose **temp1.mws**. You cannot load **temp1.ps** into Maple because it is PostScript format, which is not readable by Maple.

1.12.   Select File→Exit.

G.2 CHAPTER 2: MAPLE OVERVIEW

2.1. In Maple notation, enter **1/2** to produce $\frac{1}{2}$. To generate Standard Math, right click the input **1/2** and select the Standard Math menu item. You can also use the ☒ icon and Fo_r_mat→Con_v_ert to→_S_tandard Math menu.

2.2. Confirm with default Maple Notation with _F_ile→Preferences....

2.3. `[> 1+2+3;`

2.4. `[> 1+1; 2*2; 1+1: 2*2;`

2.5. `[> 1+1; # I am a comment`

2.6. Select the expression and press the icon ❋. Now the input is in executable Maple Notation. Press the ☒ icon next. The expression should appear as Standard Math Input.

2.7. You can convert plain text to Maple Input. Select the text and right-click. Select the menu Con_v_ert to→Maple _I_nput.

2.8. You see normally invisible characters that denote spaces (·) and paragraphs (¶). Press the same menu/icon again.

2.9. Select _I_nsert→Te_x_t and press Enter two times.

2.10. Enter **1+1**, **1−1**, and **1*2** inside separate execution groups. Either press the F4 or select _E_dit→Split or Join→_J_oin Execution Groups. Press Enter anywhere inside the joined execution group.

2.11. Use Maple's text search with _H_elp→_F_ull Text Search.... In Word(s), enter **square root** press Search. Scroll down a bit and select **sqrt**. You could also directly find help with **?sqrt** now that you know the function name.

2.12. `[> x := 2; X := 1;`

2.13. `[> x; X;`
       Maple is case-sensitive.

2.14. `[> restart; x; X;`
       Entering **restart** erased the values of x and X.

2.15. `[> y := 2*sqrt(x);`

2.16. `[> plot(y,x=0..100,title="Hello, I am a plot");`

2.17. `[> delta*iota*sigma;`

2.18. `[> x/sqrt(y+2);`

2.19. Enter **x*cos(y) − sin(y)**. Right-click the expression's Maple Output. Select Plots→Plot Builder. In the first window that appears, select these options: 3-D Plot and (x,y). Press Next. In the second window, select these options: Style: patch w/o grid, Axes: frame, Shading: z (grayscale), Title: Practice Plot. After you press Plot, Maple will output all of your options in the corresponding Maple Input: **plot3d(x*cos(y)-sin(y), x = -5 .. 5, y = -5 .. 5, style = PATCHNOGRID, shading = zgrayscale, axes = frame, title = "Practice Plot".**

G.3 CHAPTER 3: MAPLE LANGUAGE

3.1. **1** is an integer, **+** is an operator, **sin** is a name (and also a function), **and** is an operator and reserved word, **:** is punctuation, **()** are punctuation, **"Test Plot"** is a string.

3.2.   Maple does not produce an error message.

3.3.   **A := 1** is illegal because built-in operators, like **:=**, cannot have white space in between their characters.

3.4.   The expression $-1 + 2^3$ uses negation $(-)$, addition $(+)$, and exponentiation $(\wedge)$.

3.5.   `[> a:=1; b:=2; a*b; ab;`

The input **a*b** yields 2, whereas **ab** yields just *ab*.

3.6.   `[> 4 <= 5; 5 >= 4;`

Maple shows $4 \leq 5$ for both inputs.

3.7.   `[> ( a^2 ) / ( A + ( 1 / a ) );`

3.8.   You get an error message. The correct input is **sin((a+b))**.

3.9.   Entering **[1+3]/4** does not mean $\dfrac{(1+3)}{4} = 1$ in Maple! Maple reserves square brackets for other uses, such as lists.

3.10. No. Exponentiation is nonassociative. Enter **1^(2^3)** to evaluate 1^{2^3}.

3.11. You can use **`1`** as a Maple name. You cannot use the "plain" digit **1** as a name.

3.12. Enter **Success := Practice + Patience**. You could also enter **`Success := Practice + Patience`**.

3.13. The Maple names *Ira* and *ira* do not match due to case sensitivity.

3.14. Enter **type(gamma,protected)**. Maple will report true, which means that γ is indeed protected.

3.15. Maple prints a collection of every protected name it knows. Be glad that you don't have to memorize them all!

3.16. **in** and **or** are operators that are also reserved words.

3.17. Neither **in** or **D** allow assignments because **in** is reserved, and **D** is protected. However, you can actually back quote a reserved word to do something like **`in`:= 1** because the meaning of the reserved word does not change. Maple does not allow assignments of protected names, like **`D` := 1**, because they actually attempt to change the meaning of the protected names.

3.18. Expression tree for $\dfrac{1}{a + bc}$:

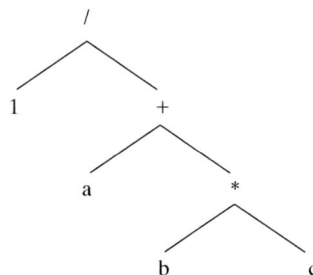

3.19. All of the given inputs are expressions.

3.20. An assignment is a task to perform. Statements do tasks.

3.21. This input builds expression by using the concept of its underlying tree.

3.22. **J=72** is an equation, and **J:=72** is an assignment. An equation does *not* assign a value!

3.23.   `[> dis := 1 + sqrt(4*b);`

3.24.   `[> restart: c := a + b: b := 1: d := c + 1;`

Maple reports **d** as $a + 2$ because **a** has no associated expression, and the value **b** is replaced with **1**, which Maple adds to the other **1**.

3.25.   `[> a; b; c; d;`

3.26.   `[> a := 'a': b:= 'b': c:= 'c': d:= 'd':`

3.27.   Use File→Preferences....

3.28.   Select Parallel Server Maple or enter **xmaple -km p**, depending on your platform. Parallel-kernel mode means that all worksheets do not "know" each other's variables.

3.29.   **%%%** corresponds to the third most recent evaluated expression, which is 3.

3.30.   Yes. Maple will usually simplify expressions, such as $x + x$ even if they are subexpressions of another expression.

3.31.   `[> A:=x+y; (2*A)/(4*A);`

3.32.   No. Maple does not know whether x is positive or negative. Enter the following:

`[> assume(x>=0); sqrt(x^2);`

Maple reports $x\sim$, where the tilde ($\sim$) indicates that x now carries an assumption.

3.33.   Maple reports these outputs in succession: B, 1, 1.

3.34.   Maple reports these outputs in succession: A, 1, 2.

3.35.   Maple reports these outputs in succession: y, y.

3.36.   `[> x:= 'x':`

3.37.   Although the forward quotes (`'`) delay evaluation, automatic simplification typically kicks in before evaluation. Therefore, the input **'1+2'** produces the value 3.

G.4   CHAPTER 4: EXPRESSION TYPES

4.1.   The surface type is **function**. Check with **type(sin(x),function)**.

4.2.   The type of $a + b$ is '+', so use **type(a+b, '+')**.

4.3.   Types **anything** and **type** correspond to a Maple expression with a recognized type.

4.4.   Automatic simplification reduces **1+2** to 3 before **whattype** checks the expression.

4.5.   Enter **72/42** to see that Maple produces $\frac{12}{7}$.

4.6.   No. Maple cannot compute integers from decimal number operations.

4.7.   `[> 123.0; 1.23e2; 1.23E2;`

4.8.   Maple produces an exact result for $123*10^{\wedge}(-1)$ because it requires only integer operations. Enter the expression using decimal points or try **evalf** as **evalf(123*10^(-1))**.

4.9.   `[> evalf(1/3,4);`

4.10.   Maple can only produce two digits, so the result must be either 2.0. or 2.1. Because rounding is set to *nearest*, Maple will round down in cases of a conflict.

4.11.   The expressions sin(0) and 10.1 are rational. The expressions $\sqrt{2}$ and *e* are irrational.

4.12.   `[> evalf(Pi,5);`

4.13.   Enter **4^(1/3)** to yield the exact form $4^{(\frac{1}{3})}$. Enter **4.0^(1/3)** to yield 1.587401052.

4.14.   No. Each input evaluates to *I*.

4.15.   Enter **(1+2*I)*(-1-I)**.

4.16.   Try dividing any expression by zero. Maple will report an error.

4.17.   `[> plot(exp(2*x), x=-infinity..infinity);`

4.18.   `[> not ((true or false) and true);`

4.19.   Enter either **solve(a*x^2+b*x+c=0,x)** or **solve(a*x^2+b*x+c,x)**.

4.20.   Use **evalb(10+2 >= 12)**, which produces true because Maple automatically simplifies **10+2**.

4.21.   No. To produce $x^2 - 1$, you need to enter **simplify((x+1)*(x-1))**. Consequently, **evalb** relies on the user to choose manipulation functions, like **simplify**, to reduce expressions before evaluating their truth.

4.22.   The statement produces the sequence 0, 1, 4, 9. Consult **?$**.

4.23.   `[> i:='i': S:=i^3 $ i=0..3;`

4.24.   Enter **S[1]** for the first element S_1. Enter **S[2]** for the second element S_2.

4.25.   `[> SL:=[S];`

4.26.   `[> SS:={S};`

4.27.   No. Enter **evalb(SS=SL)** to check. Maple distinguishes between sets and lists, even if they contain the same elements. You can, however, convert the types to each other using **convert**.

4.28.   `[> a[b] + c[d];`

4.29.   `[> constants;`

4.30.   `[> beta := 1;`

4.31.   No. Titlecase **Beta** (B) is protected, which you can discover by attempting to assign a value to it.

G.5   CHAPTER 5: FUNCTIONS

5.1.   Three arguments.

5.2.   `[> sin(theta)^2+cos(theta)^2;`

5.3.   Enter **simplify(%)** or **simplify(sin(theta)^2+cos(theta)^2)**.

5.4.   `[> y:=sin(t);`

5.5. You cannot use a trigonometric term to construct a polynomial.

5.6.
```
[> x:='x'; A:=x^2-1; B:=x^2+3x+2;
```

5.7.
```
[> C1:=A*A+A*B;
```

5.8.
```
[> C2:=A/B;
```

5.9.
```
[> expand(C1); factor(C2);
```

5.10.
```
[> r:=rem(P1,P2,x,'q'); r; q;
```

5.11.
```
[> test:=expand(P2*q)+r; evalb(test=P1);
```

5.12. Enter **roots(x^3-3*x-2)**. Maple reports $[[2, 1], [-1, 2]]$. Thus, root 2 factors the polynomial once. Root -1 factors the polynomial twice.

5.13.
```
[> R:=convert(120*degrees,radians); convert(R,degrees);
```

5.14.
```
[> sec(convert(30*degrees,radians));
```

5.15. Trick question! The $\tan\left(\dfrac{\pi}{2}\right)$ is *undefined* because you cannot divide by zero. Thus, Maple gives an error for **tan(Pi/2)**.

5.16.
```
[> convert(arcsin(0.35),degrees);
```

5.17. Enter **simplify(root(-8,3))** for the complex root $1 + i\sqrt{3}$. Enter **surd(-8,3)** for the real root -2.

5.18. Enter **log[7](163.0)** or use **evalf(log[7](163))** to find $x = 2.6176699$. Assign the result to x and enter **7^x** to check your answer.

5.19. Enter **evalf(exp(1),6)** to find $e = 2.71828$.

5.20.
```
[> assume(x>0); ln(exp(x));
```

Maple will produce x~. The natural logarithm and exponential functions are inverse functions. Try also **x:='x': invfunc[exp](x)**.

5.21.
```
[> abs(-18); abs(0); abs(18);
```

5.22.
```
[> Re(exp(I*x)) + Im(exp(I*x));
```

5.23.
```
[> S:=2^i $ i=0..3;
```

If i is already assigned, Maple will produce an error message. In that case, try either **S:=seq(2^i,i=0..3)** or **S:=2^('i') $ 'i'=0..3**.

5.24. Enter **k:='k': sum(S[k],k=1..4);** to produce $\displaystyle\sum_{k=1}^{4} S_k = S_1 + S_2 + S_3 + S_4 = 1 + 2 + 4 + 18 = 15$. Why not enter the range **k=0..3**? Because Maple sequence indices start counting from index 1, not zero.

5.25. Enter **k:='k': product(S[k],k=1..4);** to produce $\displaystyle\prod_{k=1}^{4} S_k = S_1 \times S_2 \times S_3 \times S_4 = 1 \times 2 \times 4 \times 8 = 64$. See also **?mul**.

5.26.
```
[> x:='x': dis := x -> x^3+x^2+x+1;
```

You can also enter **poly:=x^3+x^2+x+1: dis:=unapply(poly,x);**. See **?unapply** for more details.

5.27. `[> dis(-1); dis(1);`

5.28. `[> plot(dis(x),x=-1..1);`

5.29. `[> restart: f := (x,y,z) -> x+y+z;`

G.6 CHAPTER 6: MANIPULATING EXPRESSIONS

6.1. `[> convert(RootOf(x^2+1=0),radical);`

Maple evaluates $\sqrt{x^2} = \sqrt{-1}$ and determines $x = i$.

6.2. `[> convert(438,float);`

6.3. `[> simplify(onvert(sin(x),exp));`

6.4. `[> A:=sin(x)^2+cos(x)^2: combine(A); simplify(A);`
Maple produces the same result of 1 for both **combine** and **simplify**.

6.5. Enter **combine(sin(x)*sin(y))** Maple produces $\frac{1}{2}\cos(x - y) - \frac{1}{2}\cos(x + y)$, which is not a reduction. Maple converts sin to cos, when possible.

6.6. Enter **simplify(x*(x+1)-x^2)** to produce $2x^2 - x$. Try also **normal**.

6.7. Enter **normal(x+y/x)** to produce the factored normal form $\dfrac{x^2 + y}{y}$.

6.8. Enter **rationalize((x^2-1)/sqrt(x+1))** to produce $\sqrt{x + 1}(x - 1)$.

6.9. Entering **expand(factor(x^2+3*x+2))** produces the polynomial $x^2 + 3x + 2$. For polynomials, **expand** and **factor** are inverse functions.

6.10. Enter **A:=combine(sin(x)*sin(y)); expand(A);** to produce sin(x)sin(y).

6.11. **normal** does not work. Enter **expand((x+2)/(x+3)^3)** to produce $\dfrac{x}{(x + 3)^3} + 2\dfrac{1}{(x + 3)^3}$.

6.12. `[> P1:=x*y-(x^2+1)*y; P2:=collect(P1,x); P3:=sort(P2);`
This input produces the output $P1:=xy - (x^2 + 1)y$, $P2:=-y^2x + xy - y$, and $P3:=-xy^2 + xy - y$.

6.13. Enter **A:=1+sin(x+y): op(0,A); op(A);** to produce the surface type plus (+) and operands 1 and $(x + y)$. You can also enter **whattype(A)** to determine the surface type of expression A.

6.14. `[> Spring:=p=k*u; Spring;`

6.15. Maple automatically converts "greater than" ($>$) expressions into "less than" ($<$) form. Thus, entering **lhs(2>1)** produces 1, and entering **rhs(2>1)** produces 2.

G.7   CHAPTER 7: GRAPHICS

7.1.   `[> plot(x,x=0..10);`

7.2.   Activate the plot by clicking it with the left mouse button. Select <u>A</u>xes→<u>B</u>oxed from the main menu or by right-clicking the plot.

7.3.   Click the plot with the left mouse button. Point your mouse at one of the corners of the selected plot. Click with the left button again and drag the corner.

7.4.   Make sure that you have first selected the plot. Select <u>E</u>dit→<u>C</u>opy. Move the cursor to the new execution group. Then, select <u>E</u>dit→<u>P</u>aste.

7.5.   `[> plot(y^2-1,y=0..1);`

The variable y is independent, whereas the variable x is dependent.

7.6.   `[> plot(1/x,y=-10..10,discont=true);`

Without the vertical axis range, your plot might appear very flat.

7.7.   `[> f := x -> x*sin(x); plot(f(x),x=0..Pi);`

7.8.   `[> plot(exp(-x),x=-inifinity..infinity);`

7.9.   Without the vertical range **-10..10** your plot will not illustrate important features of the tangent function:

```
[> plot(tan(x), x=-Pi..Pi, -10..10, discont=true,
labels=["x","tan(x)"], title="Tangent");
```

7.10. `[> plot({x+1,-x-1},x=-2..2);`

7.11. `[> A:=plot(x+1,x=0..2): B:=plot(-x-1,x=0..2):`

7.12. Enter **with(plots)** to access **display** if you have not already done so. Next, enter **display({A,B})**.

7.13. `[> plot([cos(t),sin(t),t=0..2*Pi],scaling=CONSTRAINED);`

You drew a circle of radius 1 Why? Because $x^2 + y^2 = \cos^2 t + \sin^2 t = 1$.

7.14. `[> plot({cos(t),sin(t)},t=0..2*Pi);`

This input instructs Maple to plot both sine and cosine functions on the same plot. The previous problem plotted only *one* function that was expressed as two parametric functions.

7.15. `[> plot(exp(x^2-y^2),x=-1..1,y=-1..1);`

7.16. Enter **plot(x+y,x=-1..1,y=-1..1)**. This surface is a plane.

7.17. Enter this statement:

```
[> plot3d(log(x-y^2), x=1..1.5, y=1..1.1,
orientation=[135,45]);
```

7.18. Using a legend:

```
[> x:='x': r:=x=0..10:
[> plot([x,x^2],r,legend=["x","x^2"]);
```

Using **textplot**:

```
> with(plots):
> P1:=plot(x,r): P2:=plot(x^2,r):
> TP1:=textplot([8,15,"x"]):
> TP2:=textplot([8,80,"x^2"]):
> display({P1,P2,TP1,TP2});
```

7.19.
```
> plot([[0,0],[1,1],[2,0],[0,0]],axes=none);
```

Why enter the second "**[0,0]**"? Maple needs to know where to connect the previous point **[2,0]**.

7.20. You are plotting in cylindrical coordinates. So, enter **plot3d(z-theta^2, theta=0..360, z=-1..1, coords=cylindrical)**.

7.21. The library package **plots** contains the function **implicitplot**. If you have not already entered **with(plots)**, try **with(plots,implicitplot)** as described inside Appendix D. Then, enter **implicitplot(x^2+y^2=1, x=-1..1,y=-1..1)**. When using **implicitplot**, you should experiment with different range sizes to cover portions of the function that you might miss.

7.22.
```
> with(plottools,circle); display(circle([1,1]));
```

7.23. Enter these statements:
```
> with(plots):
> opts:=frames=10,view=-20..20,axes=none,
labels=[" "," "]:
> A:=animate(tan(x*t),x=0..100,t=0..1,opts):
> display(A,insequence=false);
```

G.8   CHAPTER 8: SUBSTITUTING, EVALUATING, AND SOLVING

8.1.
```
> restart: EQN:=y=m*x+b;
```

8.2.
```
> subs({x=1,m=2,b=0},EQN);
```

8.3. Although $\sin(0) = 0$, **subs** does not evaluate expressions. Thus, entering **subs(x=0,sin(x))** produces just $\sin(0)$. Investigate **eval** later in this chapter for evaluating substituted expressions.

8.4. Entering **subs(a*b=c,a*b*c)** will not produce c^2 because the expression ab is not an operand of abc at the surface-type level. Enter **op(a*b*c)** to check. Instead, enter **algsubs(a*b=c,a*b*c)**.

8.5.
```
> eval(x^2+cos(x),x=2);
```

8.6.
```
> eval((x^2-y^2)/(x+y),{x=2,y=3});
```

8.7. **eval** works with some modifications. Entering **eval(x*y^2*sin(x*t), x*y=1)** does not directly evaluate the subexpression because xy is not an operand of the full expression. Instead, you can trick Maple by entering **eval(x*y^2*sin(x*t),x=1/y)** to produce $y \sin (t/y)$.

8.8. Entering **solve(sin(x)=(1/2)*sqrt(3),x)** yields $\frac{1}{3}\pi$.

8.9. Entering **S:=solve(a*x^2+b*x+c=0,{x})** yields the solution set

$$S:=\left\{x = \frac{-b + \sqrt{b^2 - 4ac}}{2a}\right\}, \left\{x = \frac{-b - \sqrt{b^2 - 4ac}}{2a}\right\}.$$

8.10. `> x1:=S[1]; x2:=S[2];`

8.11. Enter **_EnvAllSolutions:=true: solve(sin(x)=0,x);** to yield $\pi\_Z1\sim$. Maple uses the name $\_Z1\sim$ to indicate the set of all integers.

8.12. Enter **fsolve(x^x=2,x)** or **solve(x^x=2.0)** to produce 1.559610470.

8.13. Enter **fsolve(sin(x)=0,x,x=10..20)**. Note that you should always check your results! Although Maple produces 15.70796327, other answers still satisfy the equation.

8.14. Entering **fsolve(sin(x)=0,x,avoid={x=0,x=Pi,x=-Pi})** produces -6.2831853.

8.15. Enter the following statements:

```
> x:='x': P:=x^2+4*x+3=0:
> S:=solve(P,{x});
> eval(P,S[1]); eval(P,S[2]);
```

Maple produces the output $S:=\{x = -3\}, \{x = -1\}$ and $0 = 0$.

8.16. Entering **map(subs,[S],P)** produces $[0 = 0, 0 = 0]$. Because you produced identities, you verified your solution set.

8.17. Entering **allvalues(RootOf(_Z^2+1))** yields the complex numbers I and $-I$.

8.18. Including complex values:

```
> S:=solve(x^5+x+1=0,{x}): evalf(S,2);
```
$\{x = -0.50 + 0.85I\}, \{x = -0.50 - 0.85I\}, \{x = -0.77\},$
$\{x = 0.87 - 0.75I\}, \{x = 0.87 + 0.75I\}$

Generate only real roots:

```
> with(RealDomain):
> evalf(solve(x^5+x+1=0,x),2);
```
$$-0.77$$

Try also **fsolve(x^5+x+1=0,{x})**.

8.19. Entering **solve(x^2>1,x)** produces the lengthy output RealRange $(-\infty, \text{Open}(-1))$, RealRange$(\text{Open}(-1), \infty)$ which translates to $(-\infty < x < -1) \cup (1 < x < \infty)$, or just the intervals $(-\infty, -1) \cup (1, \infty)$.

8.20. Enter **minimize(cos(x));maximize(cos(x));** to produce the extrema -1 and 1. Enter **minimize(tan(x));maximize(tan(x));** to produce extrema $-\infty$ and ∞.

G.9   CHAPTER 9: SYSTEMS OF EQUATIONS

9.1.   Enter the following Maple input statements:

```
> P1 := 2*x+3*y+z=0: P2:=2*y+z=21: P3:=x+z=2:
> S := solve({P1,P2,P3}, {x,y,z});
```

Maple produces the solution set $S:=\{x = 1, y = -1, z = 1\}$.

9.2.   Entering **subs(S,P1,P2,P3)** produces the verification $\{0 = 0, -1 = -1, 2 = 2\}$. Assign the solutions with **assign(S)**. Now, enter **x;y;z;** to check values of x, y, and z.

9.3.   First, enter **x:='x';y:='y':** to erase assigned values. Then, enter **solve({x+y=Pi,   sin(x)=y},{x,y})** to produce the solution set $\{x = \pi, y = 0\}$.

9.4.   The equations are linearly dependent, but you can still enter **solve ({x+y=1,2*x+2*y=2},{x,y})**. Maple reports the solution $\{y = y, x = -y + 1\}$ Thus, while you can still solve the equations, there exists no unique solution.

9.5.   Enter the following Maple input statements:

```
> with(plots);
> implicitplot({3*x+2*y=-4,-x+3*y=5}, x=-5..5, y=-5..5)
```

9.6.   Yes, you can make the attempt. Enter the following statement:

```
> implicitplot({x^2+y^2=1,y=2},x=-2..2,y=-2..2);
```

The equations do not intersect on the real plane. However, **solve** will determine an intersection with complex values. Try **complexplot3d** for graphically viewing the results.

9.7.
```
> with(LinearAlgebra);
```

9.8.   Entering **restart** removes access to the library packages.

9.9.
```
> p:=Vector([11,21]);
```

9.10.
```
> type(p,Vector); whattype(p);
```

9.11.
```
> p[1]:=10; p[2]:=20; p;
```

9.12.
```
> A:=Matrix([[1,2,3],[4,5,6]]);
```

9.13.   Matrix A has two rows and three columns.

9.14.
```
> A[1,2]; A[3,2];
```

9.15.
```
> A[1..2,2..3] := Matrix([[7,9], [11,13]]); A;
```

9.16.   Enter the following Maple input statements:

```
> M := Matrix([[1,2], [3,4]]);
> M[2,[1..2]] := Vector([a,b]);
> M;
```

9.17.   Enter **Matrix([[1,2], [2,1]]) + 2*Matrix([[-1,0], [0, -1]]) – Matrix ([[3,1], [1,3]])**.

9.18.   Enter the following statements:

```
> x:='x':y:='y':
> A:=Matrix([[1,2], [2,1]]):
> b:=Vector[row]([x,y]): c:=Vector([-1,1]):
> b . (A . c);
```

9.19.   Enter **A:=x+y=2: B:=x-y=0: solve({A,B},{x,y});** to produce the solution set $\{x = 1, y = 1\}$.

9.20.
```
> with(plots): implicitplot({A,B},x=0..2,y=0..2);
```

9.21.   Enter the following statements:

```
> with(LinearAlgebra):
> A := Matrix([[1,1],[1,-1]]):
> b := Vector([2,0]):
> x := LinearSolve(A,b);
```

See also **?LinearAlgebra[GenerateMatrix]**.

9.22.
```
> Equal(Multiply(A,x),b);
```

G.10   CHAPTER 10: INTRODUCTION TO CALCULUS

10.1.
```
> plot(tan(x),x=0..2*Pi,y=-10..10, discont=true);
```

10.2. $\tan(\pi/2)$ produces a discontinuity that might be positive of negative. Thus, Maple reports *undefined* if you enter **limit(tan(x),x=Pi/2)**. You must specify the direction to take the limit with either **limit(tan(x), x=Pi/2,right)** $(-\infty)$ or **limit(tan(x),x=Pi/2,left)** (∞).

10.3.   Enter **discont(tan(x),x)**. Maple reports $\left\{\pi\_Z2\sim + \frac{1}{2}\pi\right\}$. Maple uses _Z values as place-holders for other values and variables. In this case, variable names that initiate with _Z represent integers. Thus, values of $x = \frac{n\pi}{2}$ for $n = 1, 2, \ldots$ cause discontinuity with $\tan(x)$.

10.4.   To generate the plot, enter these statements:

```
> basicopts := xtickmarks=3, ytickmarks=3,
scaling=UNCONSTRAINED, labels =["",""]
> o1 := output = plot:
> o2:= functionoptions = [basicopts, linestyle=SOLID,
color=black, thickness=3]:
> o3 := pointoptions = [symbol=CROSS, symbolsize=200,
color=black]:
> o4 := tangentoptions = [linestyle=DOT, color=black,
thickness=3]:
> o5 := title = "Tangent to f(u) = u^2 at Point
(4,16)":
> Tangent(u^2, u=4, 0..6, o1,o2,o3,o4,o5);
```

10.5.   `[> Tangent(sin(x),x=Pi/4,output=plot);`

10.6.   `[> Tangent(a*x^2+b*x+c,x=0,output=line);`

10.7.   `[> diff(a*x^2+b*x+c,x);`

10.8.   `[> eval(diff(a*x^2+b*x+c,x),x=2);`

10.9.   `[> diff(a*x^2+b*x+c,x,x);`

10.10.   `[> diff(x*y^2+x^2*y,x);`

10.11.   Enter the following statements:

`[> ApproximateInt(x^2,x=0..4,output=value);`
`[> ApproximateInt(x^2,x=0..4,output=plot);`

10.12.   The results differ. The exact area is actually a bit larger.

10.13.   To improve the approximate, use thinner slices.

10.14.   `[> int(sin(x),x=0..Pi/2);`

10.15.   `[> int(sin(x),x);`

10.16.   Along $0 \le x \le \pi$, $\sin(x)$ is above the x-axis. However, along $\pi \le x \le 2\pi$, $\sin(x)$ is below the x-axis. Thus, entering `int(sin(x),x=0..2*Pi)` yields zero. To compute the total area, enter instead, `int(sin(x), x=0..Pi) − int(sin(x), x=Pi..2*Pi)`.

10.17.   Entering `int(k,u=0..u)` yields the function ku, whereas entering `diff(k*u,u)` yields the parameter k.

10.18.   Entering `F:=int(sin(x),x)` yields the function $-\cos(x)$. Now, you can compute the integral by entering `−cos(2*Pi) − (−cos(0))`. Maple will produce the result zero.

10.19.   `[> AntiderivativePlot(sin(x),x);`

Appendix H

Command Summary

H.1   Escape Characters and Punctuation

Symbol	Description	Example
?	Find help on topic or function	`?help`
#	Provide inert comment	`# Comment`
\	Continue input; insert control character	`A \ B;`
!	Escape to operating system. See `?escape`	`!dir`
;	Statement separator (show output)	`1+1;`
:	Statement separator (suppress output)	`1+1:`
,	Expression separator	`1, 2, 3`
	Column separator	`<1,2>`
.	Decimal point	`1.23`
"	String	`"string"`
'	Delay evaluation	`A:='A'`
`	Form symbols and names	`` `Maple rules!` ``
~	Indicate variable carries assumption; spreadsheet variable, e.g., $x\sim$	`assume(x>=0)`
( )	Separate expressions and other Maple language elements	`A*(B+C)`
[ ]	Lists, e.g., a, b, c	`L:=[a,b,c]`
	Indexing and selection, e.g., L_1, L_2, L_3	`L[1], L[2], L[3]`
{ }	Sets	`S:={a,1}`
<>	Vectors, Matrices	`<<1,3>|<2,4>>`
|	Row separator	`<1|2>`

H.2   Operators

Operator	Description	Example
%	Ditto	`%; %%; %%%`
!	Factorial, e.g., 3!	`3!`
^	Exponentiation, e.g., x^2	`x^2`
@@	Find inverse function, e.g., $f^{-1}(x)$	`f@@(-1)`

Operator	Description	Example
`*`	Multiply, e.g., $x \times y$	`x*y`
space `.` *space*	Multiply matrices, e.g., AB	`A . B`
`/`	Divide; fraction, e.g., $x \div y = \dfrac{x}{y}$	`x/y`
`intersect`	Set intersection, e.g., $A \cap B$	`A intersect B`
`+`	Addition, e.g., $x + y$	`x+y`
`−`	Subtraction, e.g., $x - y$	`x-y`
`union`	Set union, e.g., $A \cup B$	`A union B`
`minus`	Set subtraction, e.g., $A - B$	`A minus B`
`..`	Range, e.g., $a \le x \le b$	`x:=a..b`
`<`	Less than, e.g., $1 < 2$	`1 < 2`
`<=`	Less than or equal, e.g., $1 \le 2$	`1 <= 2`
`>`	Greater than, e.g., $2 > 1$	`2 > 1`
`>=`	Greater than or equal, e.g., $2 \ge 1$	`2 >= 1`
`=`	Equal; equation, e.g., $y = mx + b$	`y=m*x+b`
`<>`	Not equal, e.g., $1 \ne 2$	`1 <> 2`
`$`	Sequence, e.g., $2^i = 1, 2, 4$ for $i = 0, 1, 2$	`2^i $ i=0..2`
`not`	Logical negation, e.g., $\neg P$	`not P`
`and`	Logical "and", e.g., $P \wedge Q$	`P and Q`
`or`	Logical "or", e.g., $P \vee Q$	`P or Q`
`->`	Functional notation, e.g., $f(x) = x^2$	`f := x->x^2`
`,`	Expression sequence, e.g., $x, y, 2^a$	`x, y, 2^a`
`assuming`	Assume a property	`sqrt(x^2) assuming x>=0`
`:=`	Assignment, e.g., $x := 2$	`x := 2`

H.3 RESERVED WORDS AND STATEMENTS

Keyword	Description	Example
`description`, `options`	Label and customize a procedure	`description "this was created by me"`
`end`	End a control structure	`end`
`ERROR`	Supply an error message	`ERROR`
`for`, `from`, `by`, `to`, `while`, `do`, `end do`	Repetition statements	`for i from 1 to 10 do L[i]:=I^2; end do;`
`if`, `then`, `elif`, `else`, `end if`	Selection statements	`if type(A,even) then print ("Even!") end if;`
`local`, `global`	Declare variables	`local i, temp;`
`proc`	Procedure	`myProgram := proc(A) if(type(A, numeric) then A^2; 2*A; end if; end proc;`

Keyword	Description	Example
quit, stop, done	Exit from command-line Maple	quit
restart	Restart a Maple session without exiting	restart
save, read	Save to a file Retrieve from a file	save "work.m" read "work.m"

H.4 MISCELLANEOUS NAMES AND CONSTANTS

Name	Description	Example
constants	Print Maple constants: *false, γ, ∞, true, Catalan, FAIL, π*	constants
Digits	Number of digits Maple should use in numerical evaluations; resembles significant figures	Digits := 3
Fail	Logical "fail"	Fail
false	Logical "false", F	evalb(1=2)
gamma	Euler's constant, γ	gamma
I	Imaginary number, $\sqrt{-1} = i$	I
infinity	Infinite value, ∞	infinity
Pi	Mathematical constant Pi, π	Pi
rounding	Determine how Maple rounds numerical expressions	rounding := nearest
true	Logical "true", T	evalb(1=1)
undefined	"Value" for undefined operations	infinity-infinity

H.5 FUNCTIONS

Function	Description	Example
abs	Absolute value, e.g., $|x|$	abs(x)
algsubs	Algebraic substitution	algsubs(x*y=c,x*y*z)
alias	Abbreviate Maple names	alias(I=I,j=sqrt(-1))
allvalues	Find all solutions from a **RootOf** expression	allvalues(RootOf(_Z^2+1))
anames	Produce a sequence of currently assigned names	anames()
animate	Animate a Maple plot	plots[animate] (cos(x*t)^2, x=0..10, t=0..10)
ApproximateInt	Approximate the area underneath a function	Student[Calculus1] [ApproximateInt] (2*u,u=0..4, partition=6, output=sum);
arcsin	Inverse of sine function	arcsin(y/r)
array	Assign an array	array(0..10)
assign	Assign a name	assign(c=4)
assume	Restrict or constrain a variable	assume(x>0)

Function	Description	Example
`cat`	Concatenate, or append, strings	`cat("hi","ho")`
`circle`	Plot a circle	`plottools[circle]([0,0])`
`combine`	Combine expressions	`combine(sin(x)^2 + cos(x)^2)`
`Complex`	Create complex number	`Complex(1,3)`
`collect`	Collect like terms inside an expression	`collect(x*(x^2-x*y),x)`
`Constant`	Retrieve a constant	`ScientificConstants [Constant](g,units)`
`convert`	Convert expression types	`convert([1,2],set)`
`cos`	Cosine	`cos(Pi/4)`
`cosh`	Hyperbolic cosine	`cosh(Pi/4)`
`cot`	Cotangent	`cot(Pi/4)`
`CrossProduct`	Cross product	`LinearAlgebra [CrossProduct](a,b)`
`csc`	Cosecant	`csc(Pi/4)`
`denom`	Extract the denominator of a fraction	`denom(a/b)`
`Determinant`	Determinant of matrix, e.g., $|A|$	`LinearAlgebra [Determinant](A)`
`diff`	Differentiation	`diff(y(x),x)`
`display`	Display 2D plot structures	`plots[display]({P1,P2})`
`display3d`	Display 3D plot structures	`plots[display3d]({P1,P2})`
`DotProduct`	Dot product	`LinearAlgebra [DotProduct](v1,v2)`
`dsolve`	Solve differential equation	`dsolve(diff(f(x),x)=C)`
`Element`	Retrieve information about a chemical element	`ScientificConstants [Element] (gold,density,units)`
`Equal`	Test equality of vectors and matrices	`LinearAlgebra[Equal](A,B)`
`eval`	Evaluate an expression	`eval(x^2,x=a)`
`evalb`	Evaluate the boolean value of an expression	`evalb(1<2)`
`evalc`	Evaluate complex expressions	`evalc(2*I+I^3)`
`evalf`	Evaluate a floating-point value of an expression	`evalf(2*Pi)`
`evaln`	Evaluate to a name	`A:=evaln(A)`
`evalr`	Evaluate a range	`evalr(INTERVAL(1..2)+ INTERVAL(3..4))`
`exp`	Exponential function, e.g., e^x	`exp(x)`
`expand`	Expand an expression	`expand(x*(x+1))`
`factor`	Factor an expression	`factor(x^2+3*x+2)`
`Float`	Create a floating-point number	`Float(2,3)`
`fsolve`	Numerical solve an equation	`fsolve(x^3+17*x-3=0,x)`
`IdentityMatrix`	Produce an identity matrix	`LinearAlgebra [IdentityMatrix](3)`
`ifactor`	Find integer factors of an expression	`ifactor(72)`
`Im`	Extract the imaginary component of a complex expression	`Im(exp(2*I))`

Function	Description	Example
`implicitdiff`	Take an implicit derivative	`implicitdiff(x^2+y^2,x,y)`

`implicitplot`	Plot an implicit function	`plots[implicitplot]`
		`(x^2+y^2=1,`
		`x=-1..1,y=-1..1)`

`int`	Integrate an expression	`int(x^2,x=0..1)`
`interactive`	Activate the plot builder for an expression	`interactive(cos(x)^2+sin` `(y)^2)`
`interface`	Modify a user interface variable	`interface(verboseproc=2)`
`INTERVAL`	Create an interval, e.g., [1, 2]	`INTERVAL(1..2)`
`lhs`	Extract the left-hand side of an expression	`lhs(y=m*x+b)`
`limit`	Find the limit of a function	`limit(tan(x),x=Pi/2,left)`
`LinearSolve`	Solve a linear system of equations	`LinearAlgebra` `[LinearSolve](A,b)`
`ln`	Natural logarithm	`ln(x)`
`log`	Natural logarithm	`log(x)`
`log[]`	General logarithm	`log[10](x)`
`log10`	Base-10 logarithm	`log10(x)`
`map`	Apply a procedure to each operand in an expression	`map(sin,[0,Pi/2,Pi])`
`Matrix`	Create a matrix	`LinearAlgebra[Matrix]` `([[1,2],[2,1]])`
`MatrixAdd`	Add two matrices	`LinearAlgebra` `[MatrixAdd](A,B)`
`MatrixInverse`	Invert a matrix, e.g., A^{-1}	`LinearAlgebra` `[MatrixInverse](A)`

`MatrixScalarMultiply`	Multiply a matrix by a scalar	`LinearAlgebra`
		`[MatrixScalarMultiply]`
		`(A,c)`

`MatrixMatrixMultiply`	Multiply two matrices	`LinearAlgebra`
		`[MatrixMatrixMultiply]`
		`(A,B)`

`MatrixVectorMultiply`	Multiply a vector by a matrix	`LinearAlgebra` `[MatrixVector](A,b)`
`max`	Find maximum value from sequence	`max(1,2,3)`
`maximize`	Find maximum value of an expression	`max(sin(x))`
`min`	Find minimum value from sequence	`min(1,2,3)`
`minimize`	Find minimum value of an expression	`minimize(sin(x))`
`Multiply`	Multiply matrices	`LinearAlgebra` `[Multiply](A,B)`

`NewtonQuotient`	Find Newton Quotient of an expression	`Student[Calculus1]`
		`[NewtonQuotient]`
		`(u^2,u=4)`

`normal`	Reduce rational expressions	`normal((x^2-1)/(x+1))`
`numer`	Extract the numerator of a fraction	`numer(a,b)`
`NumericClass`	Identify a numerical type	`NumericClass(17.9)`

Function	Description	Example
`op`	Display operands inside an expression's surface type	`op(sin(x+y))`
`packages`	Display list of currently accessed library packages.	`packages()`
`piecewise`	Create a piecewise function	`piecewise(0<x,exp(x))`
`pointplot3d`	Plot points in 3D	`plots[pointplot3d]([[1,1,1], [2,2,2]], connect=true)`
`plot`	Plot an expression	`plot(x^2,x=0..1,y=1..2)`
`plot3d`	Plot a 3D expression	`plot3d(x*y,x=0..1,y=0..1)`
`plotsetup`	Customize plot display	`plots[plotsetup](ps, plotoutput='test.ps')`
`polyhedraplot`	Draw polyhedra	`plots[polyhedraplot]([0,0,0], polytype = SnubDisphenoid)`
`print`	Print a string, name, or expression	`print("Ira was here!")`
`product`	Multiply a sequence of values	`product(x[i],i=1..n)`
`quo`	Evaluate the quotient from polynomial division	`quo(a,b,x,'r')`
`rationalize`	Remove radicals from a denominator.	`rationalize(1/sqrt(2))`
`Re`	Extract the real component from a complex expression	`Re(exp(2*I))`
`RealRange`	Numeric range, e.g, $a \le x \le b$	`RealRange(40, infinity)`
`rem`	Evaluate the remainder from polynomial division	`rem(a,b,x,'q')`
`rhs`	Extract the right hand side of an equation	`rhs(y=m*x+b)`
`root`	Evaluate the nth root of an expression, e.g., $\sqrt[n]{x}$	`root(x,n)`
`RootOf`	Place-holder for roots of an expression	`RootOf(x^2+1)`
`roots`	Evaluate roots of a polynomial	`roots(x^2+3*x+2)`
`sec`	Secant	`sec(x)`
`seq`	Generate sequence of expressions	`seq(I^2,i=0..1)`
`series`	Expand expression into a series	`series(cos(x),x=0,6)`
`simplify`	Simplify an expression	`simplify(sqrt(x^2), symbolic)`
`sin`	Sine	`sin(x)`
`sinh`	Hyperbolic sine	`sinh(x)`
`smartplot`	Create a plot to tinker with	`smartplot(cos(x))`
`solve`	Solve an equation	`solve(y=m*x+b,x)`
`sort`	Sort operands of a list or polynomial	`sort([b,a])`
`sqrt`	Square root, e.g., $\sqrt{x}$	`sqrt(x)`
`subs`	Substitute expressions for operands in another expression	`subs(x=a,x*y)`
`sum`	Sum a sequence of expressions	`sum(x[i],i=1..n)`
`surd`	Find real number roots	`surd(-1,3)`

Function	Description	Example
`table`	Create a table expression	`me:=table([name1="Dave",` `name2="Schwartz"])`
`tan`	Tangent	`tan(x)`

`Tangent`	Find slope or plot tangent at a point	`Student[Calculus1]`
		`[Tangent] (x^2, x=4,`
		`output=slope);`

`textplot`	Label plots with text strings	`with(plots):` `textplot([0,1,"Hello"])`
`Transpose`	Transpose a vector or matrix	`with(): Transpose(A)`
`type`	Check an expression's type	`type(1,integer)`
`unames`	Displace sequence of unassigned, but used, names	`unames()`
`unassign`	Remove a variable's assignment	`x:=2: unassign('x')`
`unprotect`	Remove a protected name's protection	`unprotect(gamma)`
`UseSystem`	Set a default system of units	`Units[UseSystem](FPS):` `with(Units[Natural]):`
`value`	Evaluate inert functions	`value(Int(x^2,x))`
`Vector`	Create a vector	`LinearAlgebra` `[Vector([1,2])`
`VectorAdd`	Add two vectors	`LinearAlgebra` `[VectorAdd(u,v)]`
`VectorAngle`	Find angle between two vectors	`LinearAlgebra` `[VectorAngle(u,v)]`
`VectorNorm`	Find a vector norm, like magnitude	`LinearAlgebra` `[VectorNorm(v,2)]`

`VectorScalarMultiply`	Multiply a vector by a scalar	`LinearAlgebra`
		`[VectorScalarMultiply]`
		`(v,c)`

`whattype`	Determine an expression's type	`whattype(sin(x))`
`with`	Load library package	`with(LinearAlgebra)`

Index